Agroecosystems – Very Complex Environmental Systems

Edited by Marcelo L. Larramendy and Sonia Soloneski

Published in London, United Kingdom

IntechOpen

Supporting open minds since 2005

Agroecosystems – Very Complex Environmental Systems
http://dx.doi.org/10.5772/intechopen.87489
Edited by Marcelo L. Larramendy and Sonia Soloneski

Contributors
Shalika Mehra, Carmine Massarelli, Claudia Campanale, Vito Felice Felice Uricchio, Ayşe Ezgi Ünlü, Serpil Takaç, Sadhana Balasubramanian, Bagampriyal Selvaraj, Zulias Mardinata Zulkarnaini, Mardaleni, Tengku Edy Sabli, Pavel Prudnikov, Zoya Ozherelieva, Diana Krivushina, Marina Zubkova, Anna Androsova, A. K. M. Mominul Islam, A. K. M. Aminul Islam, Swapan Chakrabarty, Zahira Yaakub

First published in London, United Kingdom, 2021 by IntechOpen
IntechOpen is the global imprint of INTECHOPEN LIMITED, registered in England and Wales, registration number: 11086078, 5 Princes Gate Court, London, SW7 2QJ, United Kingdom
Printed in Croatia

British Library Cataloguing-in-Publication Data
A catalogue record for this book is available from the British Library

Additional hard and PDF copies can be obtained from orders@intechopen.com

Agroecosystems – Very Complex Environmental Systems
Edited by Marcelo L. Larramendy and Sonia Soloneski
p. cm.
Print ISBN 978-1-83880-384-1
Online ISBN 978-1-83880-528-9
eBook (PDF) ISBN 978-1-83880-529-6

For EU product safety concerns:
IN TECH d.o.o., Prolaz Marije Krucifikse Kozulić 3, 51000 Rijeka, Croatia, info@intechopen.com or visit our website at intechopen.com.

Meet the editors

Marcelo L. Larramendy, Ph.D., serves as a Professor of Molecular Cell Biology at the School of Natural Sciences and Museum (National University of La Plata, Argentina). He is the appointed Senior Researcher of the National Scientific and Technological Research Council of Argentina and a former member of the Executive Committee of the Latin American Association of Environmental Mutagensis, Teratogenesis and Carcinogenesis. He is the author of more than 450 contributions, including scientific publications, research communications, and conferences worldwide. He is the recipient of several national and international awards. Prof. Larramendy is a regular lecturer at the international A. Hollaender courses organized by the IAEMS and a former guest scientist at NIH (USA) and University of Helsinki (Finland). He is an expert in genetic toxicology and is, or has been, a referee for more than 20 international scientific journals. He was a member of the International Panel of Experts at the International Agency for Research on Cancer (IARC, WHO, Lyon, France) in 2015 for the evaluation of DDT, 2,4-D and Lindane. Presently, Prof. Dr. Larramendy is Head of the Laboratory of Molecular Cytogenetics and Genotoxicology at the UNLP.

Sonia Soloneski has a Ph.D. in Natural Sciences and is an Assistant Professor of Molecular Cell Biology at the School of Natural Sciences and Museum of La Plata, National University of La Plata, Argentina. She is a member of the National Scientific and Technological Research Council (CONICET) of Argentina in the genetic toxicology field, the Latin American Association of Environmental Mutagenesis, Teratogenesis and Carcinogenesis (ALAMCTA), the Argentinean Society of Toxicology (ATA), the Argentinean Society of Genetics (SAG), the Argentinean Society of Biology (SAB), and the Society of Environmental Toxicology and Chemistry (SETAC). She has authored more than 380 contributions in the field, including scientific publications in peer-reviewed journals and research communications. She has served as a review member for more than 30 scientific international journals. She has been a plenary speaker in scientific conferences and a member of scientific committees. She is a specialist in issues related to genetic toxicology, mutagenesis, and ecotoxicology.

Contents

Preface

It is currently well known and widely accepted that farmers are the general managers of the Earth's land surface worldwide. Furthermore, they will continue to shape agricultural environments in the coming decades. The generation of increased and improved innovative methodologies and approaches that guarantee the sustainability of crop production, simultaneously with ecosystem services, is a challenging scientific area, largely led by ecologists, agronomists, and theoreticians, who must address this task as a joint effort. Agricultural ecosystems offer a variety of benefits to the global population, and these are recognized as Ecosystem Services. In this context, it has been acknowledged that ecosystem services are both actively or passively engaged in enhancing the well-being of the global population. Accordingly, as defined in the Millennium Ecosystem Assessment (2005), the main role of agricultural practices is to support provisioning ecosystem services, mainly through the manufacture of supplies and educational management. The ecological tools underlying ecosystem services take into account the complexity of the many possible relationships between species of economic importance and the characteristics of local ecosystems. Furthermore, the consequences of biodiversity on the mechanisms underlying most ecosystem services are nowadays well recognized and known in most cases, both empirically and theoretically. In an ideal condition, most, if not all, ecosystem services should also be considered when resource management decisions are taken and should be included at different scales, both spatial and temporal. It is also important to avoid overlooking the interactions between biodiversity and stakeholders, in order to maximize the benefits derived from such practices and minimize the cost through appropriate decisions. However, relationships between agricultural practices and ecosystem services are, in most cases, not totally understood. The ecological mechanisms underlying ecosystem services include multifaceted interactions not only between organisms, but also among different types of organisms, and cultural practices and stakeholder strategies can either inhibit or enhance the ecosystem services. In reality, mechanistic modeling tools examining the consequences of management options on the provisions of most of the various ecosystem services are still lacking. There is a need for innovative cropping systems to be designed at field and landscape scales, and an even larger scales, to aid future planning, organization, and strategies, in which the complexity of the interactions encompassed by ecological and decisional networks must be included.

Today, it is accepted worldwide that an AGROECOSYSTEM represents a very complex environmental system in which many species interact, driving a variety of ecological processes at different spatial scales. In addition, agroecosystems are characterized by strong and interrelated interactions among ecological and soil management processes. These interactions encompass, in a general conceptual framework, the relationships between ecological and food webs, hosts and pathogens, and those involving spatial interdependence between localities. Under this scenario, an "interaction networks" model enables researchers in the field to (1) analyze and understand the emergent properties of complex systems, and (2) develop universal rules that allow individual stakeholders to make decisions regarding the outcomes of the ecosystem.

This book, "*Agroecosystems – Very Complex Environmental Systems*," aims to present an update on different aspects associated with the importance of sustainable agriculture. It was our intention to gather information from diverse sources in this volume and to give some real-life examples, extending the appreciation of the complexity of this subject in a way that may stimulate new approaches in relevant fields.

This book includes seven general chapters highlighting different aspects of agroecosystems worldwide. The first chapter describes the fungal endophytes of Australian orchid species; these endophytes exploit large areas of the soil, to which orchid roots have no access, and acquire both organic and inorganic nutrients beyond the depletion zone at low carbon cost. Several integrated approaches have been developed for the conservation, management, and restoration of these terrestrial orchids in the wild because appropriate conservation priorities need to be established urgently to prevent the loss of habitats for these endangered species. This chapter also focuses on the protection of these endangered Australian orchid species by developing an understanding of the nutritional behaviour of their endophytes. The second chapter aims to describe the impact of the plant hormone brassinolide on two varieties of fig from Indonesia and Malaysia, stressing the significant effect of interactions between brassinolide and diversity on fig growth and physiological changes, except in respect of plant height and dry biomass. The third chapter documents the relationship between resistant varieties of hybrid strawberries *Fragaria × ananassa* Duch. and negative environmental conditions. These conditions include physiological and biochemical indicators of resistance during autumn hardening and after temperature stress in winter that resulted in changes in the antioxidant system, interruptions of the protein-carbohydrate complex, accumulation of membrane lipoperoxidation products, and changes in the fractional composition of water in the leaves. The fourth chapter has been written to provide botanical descriptions of the castor bean or castor oil plant, belonging to the monotypic genus *Ricinus*, describing its ecology, agro-technology, and many industrial uses. At present the plant is in increasing demand in the international market for its more than 700 uses, ranging from medicine and cosmetics to biodiesel, plastics, and lubricants. The fifth chapter is focused on the development of a defined, highly-reliable, and integrated methodology for identifying the causes of contamination of agroecosystems in southern Italy, namely asbestos and illegal burial of waste, in the soil as well as microplastic pollution. This chapter also discusses innovative and high-speed approaches to obtaining ever more precise data on environmental degradation. The sixth chapter is an update on the literature regarding the use of deep eutectic solvents to treat lignocellulosic wastes within the field of biomass valorization. Therefore, this chapter emphasizes how the preparation of novel deep eutectic solvents and improving treatment conditions will help to solve the environmental problems originating from agro-industrial wastes and also to develop new platforms for the production of valuable products such as chemicals, biofuels, and bioactive phenolic compounds. Finally, this book includes a chapter that discusses the effects of the application of blue-green algae, which enhances the morphological and photosynthetic efficiency of the rice plant under greenhouse conditions, stressing that the application of such a bio-mixture in agriculture not only increases crop yield but also maintains our environment sustainably.

Finally, as indicated in a book we published some years ago, entitled "*Organic Fertilizers – From Basic Concepts to Applied Outcomes*", it seems obvious "that future agricultural practices will irreversibly shape the Earth's land surface, including its species, geochemistry, and disponibility of surface to the people living on it". We hope that the information presented in this book will be of value to those directly engaged in the management and use of agroecosystems, and that this book will continue to meet the expectations and needs of all those interested in the different ways that agroecosystems can be directed to achieve sustainable agriculture without compromising environmental integrity.

The chapters provided by the authors in this field of research are gratefully acknowledged. The publication of this book is orientated to those researchers, scientists, engineers, teachers, graduate students, agricultural agronomists, farmers, and crop producers who can use these different results to develop an understanding of the complexity of an agroecosystem and the different aspects and relationships among the different entities involved. The concepts of agroecosystems and ecosystem services can help scientists determine how much of each service is provided throughout the many scales of the networks (field, farm, and/or landscape) at the different ecological levels (individuals, species, communities, and ecosystems), allowing innovative strategies to be developed in ecosystem services management and the damage caused by agroecosystems to be minimized.

Marcelo L. Larramendy Ph.D. and Sonia Soloneski Ph.D.
School of Natural Sciences and Museum,
National University of La Plata,
La Plata, Argentina

Chapter 1

Fungal Endophytes: Australian Terrestrial Orchids

Shalika Mehra

Abstract

Orchids are unique as they lack a functional rooting system and share an obligate relationship with their fungal symbionts. This relationship supports their host's nutritional demands from seed germination to its later development. The orchid fungal endophytes explore large areas in the soil as, to which orchid roots have no access, and thus acquire both organic and inorganic nutrients beyond the depletion zone at low carbon cost. Both 'autotrophic' (green) and 'mycoheterotrophic' species occur in the Orchidaceae, but the term 'mixotrophic' is possibly a truer description of the carbon economy of many green orchids. Some of the major ecological threats of an Australian landscape are habitat destruction and fragmentation. There is little known about the nutritional sources and saprophytic ability of orchid mycorrhizal fungi (OMF) and their role in providing nutrition to orchids. However, several integrated approaches have been developed for the conservation, management and restoration of these plants in wild but there is an urgent need to set appropriate conservation priorities to prevent the loss of habitats for these endangered species in terms of their fungal endophytes. This chapter focuses on the protection of these endangered Australian orchid species by understanding the nutritional behavior of their endophytes.

Keywords: orchid mycorrhizal fungi (OMF), autotrophic, endangered, conservation, mycoheterotrophic (MH)

1. Introduction

Orchids (family Orchidaceae) being iconic are at the front line of extinction, with 17,000–35,000 species distributed globally and are under threat [1–3]. The family is cosmopolitan in its distribution, but the genera and species are highly endemic [4]. In the Orchidaceae, greater levels of ecological specializations associated with global climate change, have a direct impact on the species diversity and levels of threat, to the extent that many terrestrial orchids in temperate regions have become extinct.

Australia is rich in terrestrial orchid diversity (82%) with approximately 115 genera. The Southwest Australia Floristic region (SWAFR) is among 25 hotspots of biodiversity globally [5]. They can be found in a wide range of habitats across the continent and are usually categorized as epiphytes, lithophytes, and terrestrials, where epiphytes and lithophytes are mostly distributed in the warm and moist regions of tropics (18%) while few species are found in temperate regions of eastern Victoria and Tasmania [5]. They are mostly found in sclerophyll open forests and swampy coastal scrub lands. They grow on the ground especially in open habitats such as grasslands, heathlands and forest floors with low annual rainfall, showing seasonal changes and are mostly distributed in the southern temperate zones of Australia which have a Mediterranean climate. Most of the orchids growing in these temperate regions are deciduous, surviving climate extremes beneath the soil surface by undergoing dormancy [6].

They usually have subterranean fleshy thick tubers or tuberoids that store nutrients during dormancy. Some of the most common terrestrial orchid genera found in Australia are, *Caladenia* (Spider orchids), *Pterostylis* (Greenhoods), *Diuris* (Donkey orchid), *Acianthus* (Mosquito orchid), *Prasophyllum*, *Thelymitra* (Sun orchids), *Microtis* and *Glossodia* (**Figure 1**) [5]. Caladenia's are (spider orchids) endemic to Australia and represent one of the extraordinary terrestrial orchids with a large number of threatened and rare taxa [6]. In total there are 132 species of spider orchids which are mostly distributed throughout southern Australia.

From the ecological point of view, these orchids could act as ecological indicators of a healthy environment [7]. Due to their complex interactions with pollinators, fungal endophytes, and associated host trees, their conservation involves challenges at species-specific levels. These challenges are mostly linked to their habitat destruction and fragmentation, land use, climate change and unsustainable exploitation of biodiversity [8, 9]. Also, most of the terrestrial orchids of Australia, are continuously encountered by inappropriate fire regimes at different developmental stages of its life cycle, and places 74% of threatened orchid species at risk

Figure 1.
Australian terrestrial orchid species, (a) Caladenia *spp. (b)* Pterostylis *spp. (c)* Glossodia *spp. (d)* Corybas *spp. (e)* Diuris *spp. (f)* Acianthus *spp.*

of extinction [10–12]. Recently, the impact of nature-based tourism has also been reported as a major threat to the decline of threatened orchid populations in the wild in South Australia [10]. Due to these factors, the survival of various species from this genus is at risk and thus considerable effort is required from scientists and conservation practitioners to overcome these challenges of the twenty-first century. However, with the ability to use current novel technologies in orchid biology greater than ever before, we can help them conserve for future generations.

Australian soils are generally deficient in nutrients, which have mostly leached out of the sandy soils (podzols) over many millions of years [13]. In a fire-prone Australian ecosystem, fungi can have a major influence on surrounding biota and play an essential role in maintaining the healthy ecosystems as effective symbiotic partners, decomposers, nutrient cyclers and are a source of food for various organisms. The top horizon of organic matter is the major source of carbon (C), nitrogen (N) and phosphorus (P). In most coarse-rooted plants like orchids, with a poorly developed root system, mineral nutrition is highly dependent on mycorrhizal uptake of essential elements such as N and P from their surroundings [14]. Orchid mycorrhizal fungi (OMF) present in these nutrient-depleted soils are likely to derive their nutrition from the organic matter (dead roots, exoskeletons, leaves and wood in a litter), which holds various types of complex compounds. These complex molecules are further degraded into simpler forms by the activity of these mycorrhizal fungi and other microorganisms. C is usually available in complex forms such as cellulose, hemicelluloses, pectin and lignin, as well as simple soluble breakdown products from these complex polymers. Also, availability of N is usually in the form of organic peptides, proteins and amino acids and as inorganic ammonium and nitrate ions whereas phosphorus is mostly available as organic compounds such as phytic acid and sparsely available as inorganic ions such a PO_4^{3-}, HPO_4^{2-} and H_2PO_4.

The ability of OMF to assimilate various C, N and P compounds as compared to other ericoid mycorrhizal (ERM) and ectomycorrhizal (ECM) fungi, has been studied previously but information available until now is fragmentary [15–18]. It is very important to understand the nutritional physiology of endophytes associated with terrestrial orchid species while considering any recovery plans for propagation, management, conservation and restoration of Australian endangered orchid species in wild. Therefore, in this chapter, we have discussed in general about orchid endophytes and their saprophytic ability in digesting complex resources, confined to its litter prone, open and well-drained podzol sites.

2. Orchid fungal endophytes

All orchids share obligate relationships with their endophytes, from early seed germination stages to later development of seedlings and mature plants. Endophytes are commonly found inside the healthy tissues of orchid roots as bacterial and fungal endophytes without causing any symptoms of a disease. In this mutualism, fungus provides water and mineral nutrition to the host plant which in turn provides photosynthetically fixed carbon back to its fungal partner [4], phenomenon which is commonly found in fully autotrophic orchid species [19] as compared to completely mycoheterotrophic (MH) and partially MH orchid(Mixotrophic)species [20–22].

Physiology of orchid seed germination is one of the interesting phenomena of nature and therefore must enter symbiotic interaction with a species-specific symbiont for appropriate germination. All orchid species are MH in their early stages of seed development, where orchids obtain their nutrition in the form of minerals, salts, water and carbon supply from their fungal symbionts at least in their initial seed germination stages [23]. Once the fungus invades the minute orchid seeds (having low endosperm reserves) it kicks starts the germination process, eventually giving rise to an undifferentiated mass of cells known as protocorms. However, this mutual symbiosis between the host and its fungal partner has not been understood completely, it seems that orchid is having a complete control over-regulating the degree and level of these associations. Germination and vegetative propagation in their natural environment is very slow with a rate of <5% [24]. The distribution of orchids and their diversity is dependent on the availability of their fungal symbionts and thus understanding orchid mycorrhizal symbiosis is a key factor to conserve orchids.

2.1 Orchid symbionts: rhizoctonias and other mycorrhizal fungi

Various orchid species have heterobasidiomycetes as their symbionts [25]. The complex assemblage of fungi associated with orchids consists of Agaricomycetes (=Hymenomycetes) taxa [26]. OMF was traditionally classified as anamorphic form-genus (imperfect stage) *Rhizoctonia* (=*Epulorhiza*). These correspond to three distantly related basidiomycetous lineages forming teleomorphic genera, including Ceratobasidiaceae, Tulasnellaceae and Serendipitaceae [27]. Although the OMF is well known for its saprophytic abilities [4] they may be found widely as endophytes in non-orchid roots [28] without forming any symptoms of infection.

Recently, a range of mycorrhizal fungi has been found associated with different orchid species, apart from their long evolutionary history of associations with rhizoctonias [26]. OMF studies on MH and mixotrophic orchid species have shown a huge diversity of ectomycorrhizal fungi [23], including saprotrophic fungi from Mycenaceae and Psathyrellaceae and some ascomycete taxa, which suggests that depending upon their host, same fungi could have a potential to form dual associations in nature. Photosynthetic orchids can also associate with a variety of taxa, including Psathyrellaceae and saprotrophic fungal species [29].

Members of Tulasnellaceae, Serendipitaceae (Sebacinales clade B) and Ceratobasidiaceae are well known for their endophytic [30] and saprophytic abilities [31] with few exceptions from Ceratobasidiaceae where some species are plant-parasitic [27]. *Serendipita indica* is one of the well-studied, root endophyte models and is indeed found mycorrhizal with orchid roots [32]. Fungi in the Serendipitaceae are involved in a wide range of mycorrhizal associations such as ectomycorrhizas, ericoid mycorrhizas, orchid mycorrhizas and even liverworts (Jungermannioid mycorrhizas) [26, 33–35]. Phylogenetically the Serendipitaceae (formerly called Order Sebacinales) is grouped into two clades: A and B [36]. Clade A species constitutes jelly fungi, having a saprophytic ability through which they can obtain their nutritional demands from wood and other surrounding litter present in their habitat, while Clade B species are common endophytes of underground plant organs [37]. The fungi from Clade B are usually associated with orchids, for example, *Caladenia* species in Australia, are also associated with ericoid roots, though without having any proof of functional symbiosis so far [33]. There are studies which have shown presence of basidiomycetous hyphae with septal pores on and in sections of ericoid plants by transmission electron microscopy (TEM) whereas, there is an evidence of DNA sequences from the nuclear ribosomal internal transcribed spacer (ITS) region from ericoid roots that grouped within Serendipita group B and contained identical sequences to those from *Serendipita vermifera* isolates from Australian green orchids [4]. *S. vermifera* [30] in group B [38], has a confirmed mycorrhizal relationship with some green orchids, e.g. in *Caladenia* and *Glossodia* species and is the most common OMF found associated to these taxa [39–42].

2.1.1 Fungal identification

Traditional approaches were commonly used to identify these fungal endophytes of orchids by isolating the pelotons from the orchid tissues and maintaining them as pure cultures. Mycelia are mostly present as anamorphs and the orchid endophytes are commonly identified based on their morphological (hyphal walls), anatomical differences (spore formation and nucleus number) and anastomosis behavior [43] by using optical, scanning and electron microscopy. Most form chains of small ovoid-globular monilioid cells. Recently, several molecular approaches are extensively used to delimit the fungal endophytes of orchids (*Ceratobasidium*, *Tulasnella* and *Rhizoctonia* = *Serendipita*) which are well known for their poor taxonomy [44, 45].

2.1.1.1 Asexual stages

Rhizoctonia is remarkable in some characteristics as they branch out at acute angles when young but at right angles to the main axis at maturity, mainly constricting at the point of branching [46]. Fungi grown from pelotons usually form ovoid monilioid cells without having any clamp connections or conidia in a culture that limits their identification through morphological methods [40]. *Rhizoctonia*, is traditionally characterized on the basis of anastomosis groupings including the pathogenic strains [47]. *Rhizoctonia* species are separated on the basis of the ultrastructure of the number of nuclei in each cell and the septa, and on the basis of what can be categorized as uninucleate, binucleate or multinucleate [48]. The commonly isolated *Rhizoctonia* fungi from terrestrial orchid species are within the anamorphic genera *Ceratorhiza, Moniliopsis, Thanatephorus* and *Epulorhiza*.

2.1.1.2 Sexual stages

The commonly isolated *Rhizoctonia* fungi from terrestrial orchids are species within the teleomorphic genera, *Ceratobasidium, Tulasnella* and *Serendipita*. Imperfect stages of *Rhizoctonia* are commonly found in various chlorophyllous orchids. For most Australian green orchids, *in-vitro* cultures produce only monilioid cells but Warcup and Talbot obtained teleomorphic stages on *Rhizoctonia* isolates in culture [40, 41, 49], an achievement not replicated by many researchers despite numerous attempts. Because of this, the systematics of *Rhizoctonia*-type OMF has been studied using both morphological [43, 46] and molecular approaches [25, 38, 44, 45, 50], which have suggested various anamorphs and teleomorphs for this polyphyletic group.

3. Fungal endophytes of myco-heterotrophic (MH) and autotrophic orchids

The nutrition of orchids is closely tied to the nutrition of their basidiomycetous OMF. The fungal symbionts provide essential nutrients for the establishment of orchid seedlings from obligate MH stage to mixotrophic to fully autotrophic stages of their development. They can obtain their nutrition as saprophytes, by breaking down wood and other litter in their habitats or by tripartite symbiosis, in which the OMF is also ectomycorrhizal on the roots of the surrounding higher plants. Both result in networks of hyphae linking the host plants to various habitats.

In general, achlorophyllous orchids mostly have mycorrhizal associations with homobasidiomycete fungi in the Cantharellales, Thelephorales, Agaricales, Serendipitaceae, Hymenochaetales, and Russulales, which are also pathogenic and ectomycorrhizal on higher plants [51]. In MH orchids, the fungi often form tripartite relationships, being ectomycorrhizal with woody plants and endomycorrhizal with orchids [23, 52, 53] where, transfer of carbon has been shown from the woody plants to the orchid [52, 54]. Fungal symbionts of MH orchids have three lifestyles: ectomycorrhizal (ECM), e.g. *Corallorhiza*-Russulaceae, parasitic (pathogenic), e.g. *Gastrodia—Armillaria* species, and saprophytic, e.g. *Epipogium—Coprinus* and *Psathyrella* species. Various achlorophyllous orchids such as *Gastrodia confusa* [55], *G. elata* [56], *Epipogium roseum* [57] and *Fulophis zollingeri* [58], are associated with many species of saprophytic wood- and litter- decaying fungi. Earlier studies have provided morphological and ultrastructural evidence that fungi from the Serendipitaceae formed ectomycorrhiza with *Corylus avellana* and *Carpinus betulus* [25] suggesting that common mycorrhizal networks (CMNs) are likely to be found in the plant communities where MH orchids are distributed in the close vicinity of ectomycorrhizal higher plants where they can obtain their nutrition through a tripartite relationship. Molecular studies have also shown the presence of *Serendipita* species on MH orchids such as *Hexalectris spicata* and *Neottia nidus avis*, suggesting that, if *Serendipita* is ubiquitous in its distribution, it is of interest to elucidate any functional symbiosis with ECM on higher plants.

Chlorophyllous orchids mostly have mycorrhizal associations with fungi in the *Rhizoctonia* alliance, in the Cantharellales and Sebacinales (*Serendipita* Group B), with sexual stages in the Ceratobasidiaceae, Serendipitaceae and Tulasnellaceae [4]. Some of the *Rhizoctonia* species in the Ceratobasidiaceae are also plant pathogens of crops [4]. Fungal endophytes from the *Serendipita* group are common among photosynthetic orchids, e.g. *Caladenia* [42, 59] and non-photosynthetic terrestrial orchids, e.g. *Neottia* [53, 60, 61]. They constitute two major groups: A and B [36]. Group B forms mycorrhizae with green orchids while group A is generally associated with ECM and some non-photosynthetic orchids [26].

4. Fungal specificity

Fungal specificity is common in Australian terrestrial orchids [39, 62]. Taxonomically related groups of Australian terrestrial orchid genera are associated with taxonomically related groups of fungi. Both achlorophyllous and chlorophyllous orchid species can have fungal specificity [57, 63] but is more remarkable among heterotrophic orchid species [64]. By contrast, chlorophyllous photosynthetic mycorrhizal plants are said to be generalists in their associations with mycorrhizal fungi [4], though there is evidence of specificity at the species and strain level in Australian OMF and their host orchids, especially *Caladenia* [17, 65].

Most common genera of seasonally dormant terrestrial orchids in Australia belong to the Tribe Diurideae; within this, genera in the Sub-tribe Prasophyllinae usually associate with *Ceratobasidium*, those in the Caladeniinae with *Serendipita*, and most of those in the Diuridinae, Drakaeinae and Thelymitrinae associate with *Tulasnella*. Genera in the Acianthinae and the Megastylidinae associate with *Serendipita* and/or *Tulasnella*, e.g. *Thelymitra. calospora* and *Lyperanthus nigricans* associated with a wide range of endophytes. Also, variations in seed germination rates with fungal isolates of *T. calospora* were noticed in *Diuris* species [39]. However, within these general relationships, fungal strain, seed and fungal provenance play an important role; specificity varies from high in *C. tentaculata*, in which seed and fungal provenance both varied seed germination significantly, to low, in which more than one species of *Tulasnella* stimulated germination in *Thelymitra* [39].

OMF effectiveness leads to increased seed germination rate and fitness of orchids [66]. Specificity can be strictly restricted to the early seed germination stages of orchid or involve the compatibility of the fungal symbiont with the orchid throughout later stages [67]. Masuhara and Katsuya [62] has expanded fungal specificity into "potential and ecological specificity" whereas, earlier *in-situ* seed baiting studies from endangered and common orchids have shown distributions of OMF independent of their host orchids [68], suggesting that the patchiness of many orchids is not due to patchiness of their compatible species.

Previous research has also shown fungal specificity with particular orchid species during germination stages; for example, *Neottia nidus-avis* needs a specific *Serendipita*-like fungus to germinate [61]. Fungal specificity and effectiveness vary with individual isolates associated with the host orchid species for example, OMF isolated from *Caladenia* species were effective in germinating seeds of both *Caladenia* and *Glossodia* as compared to *Eriochilus cucullatus* and *Acianthus reniformis* [39]. These seed germination tests, under in-vitro conditions, over-estimate the potential of OMF isolates to form effective symbioses with orchid species, and results in a failure of symbiosis during later stages of orchid development thereby parasitizing the host plant [69]. Also, it does not explain the fungal switching that has been recorded during the lifetime of an orchid in the wild [70].

5. Nutritional trends in OMF

Decomposition of organic materials present in the form of dead decaying material such as fallen leaves, litter, hair, exoskeletons and any other kind of waste product from plants or animals is the main source of carbon compounds available.

In forest ecosystems, mineral nutrients in the form of P and N are mostly locked within living organisms or in the organic layer of soil. The distribution of these resources is heterogeneous in terms of space and time [71]. Access to nutrients by the host plant depends on the ability of the mycorrhizal fungi to mineralize the available organic nutrients to intermediate and soluble forms and then mobilize them to the host plant [72]. OMF can grow freely in the environment and have an ability to sustain itself without its host [21]. Mycelium is the predominant vegetative form among the basidiomycetes, comprising interconnected hyphae [71]. Fungal foraging for the uptake of minerals and other resources that are interlocked in the organic layer of the soil largely takes place at hyphal tips. Fungal hyphae have a large surface to volume ratios and secrete enzymes that digest extracellular organic resources, which are further translocated to a sink in the form of simple soluble compounds [73]. From the nutrient-deprived ecosystems of Australia, very limited information is available on the ability of OMF to utilize various C, N and P sources from the complex litter present on the forest floors.

For successful symbiotic interactions, efficient utilization of nutrients by the fungal partners is a prerequisite. In most mycorrhizal associations, photosynthetic products are transferred from an autotrophic host plant to a heterotrophic fungal partner, while the mineral nutrients obtained from the soil move in the opposite direction [74]. By contrast, in mycorrhizae of the photosynthetic orchids, the flow of nutrients is bidirectional, at least in some orchids [19]. In orchids, nutrient uptake into OMF occurs mainly through the acquisition of soluble nutrients from the decay of organic litter present in the top 4–12 cm of topsoil [73]. Information on the types of soluble carbon sources OMF can utilize from the environment and their host plants are very limited.

Few studies have reported inter- and intra- specific variations in utilization of substrates among orchid and ericoid mycorrhizal fungi from the same habitat [16, 18]. Also, Wright et al. [17] provided evidence of genetic and functional diversity among OMF isolates of *C. tentaculata* that varied in germination rates and utilization of some C and N sources. Unlike many ECM basidiomycetes, OMF has also retained the genes for the breakdown of these complex carbon compounds [31]. Understanding the nutritional roles of OMF may explain the diversity noticed among fungal isolates, from even single orchid plants in rates of symbiotic seed germination *in vitro*. However, in most cases, only one symbiotically effective fungus was examined from each orchid species from their habitat despite, a large number of fungal variations commonly isolated from even single plants. The symbiotic effectiveness of these isolates might vary with their ability to take up and utilize various carbon sources from their surroundings, an aspect that has not been studied so far.

5.1 Carbon sources: saprophytic ability of orchids and their dependence on mycorrhizal partners

During the early stages of orchid seed development, both achlorophyllous and fully autotrophic orchid species lack their ability to synthesize carbohydrates and the only available source of carbon and nitrogen to these plants is through OMF associated to them. One of the common assumptions so far in the orchid biology is that OMF can obtain its nutrition by digesting the litter components present on the forest floors and there has not been much evidence of their ability to grow on these litter components apart from few studies [15–18, 75]. There are reports where orchids are found in close vicinity of moss lying on the forest floors but there is no scientific evidence showing the presence of OMF on them or surrounding litter [15]. *S. vermifera* complex is mostly root biotrophic [37] and is associated with *Caladenia* species that is believed to be saprotrophic, at least as far as the fungi isolated from the Australian orchids is concerned.

In their natural habitat's orchids are commonly surrounded by litter such as bark, leaves and wood. During *ex-situ* measures for orchid conservation, these components have been extensively used as mulch in the pots of orchids to retain proper moisture levels. In Australia, Casuarina branchlets are commonly used as a source of mulch for re-emergence and growth of orchids during *ex-situ* conservation measures based on an assumption that they help orchid leaves from drying up but there is a possibility that these litter components on their break down may help them in the nutrition of the OMF and hence the orchid growth [75]. Recently, Mehra et al. [15] have validated their use in *ex-situ* cultivations by showing the amounts of fungal biomass produced on natural and semi-purified substrates from various endangered and common *Caladenia* species under *in-vitro* conditions.

5.1.1 *Complex carbon sources in a litter*

Nutrient-poor soils are inadequate in their microbial decomposition rates and the dead organic matter present on the soil is mostly utilized by decomposer fungi [76]. Litter constituting bark, wood, and leaves have biopolymers such as chitin, pectin, lignin, cellulose, hemicellulose and contain complex cell wall polysaccharides along with chitin of fungal and invertebrate origin. Most of this organic waste is in the form of plant cell wall components which constitutes 90% of plant cell wall components, having three major polysaccharides: cellulose, hemicelluloses and pectin [77]. Of these, cellulose and pectin are key components of organic substrates in vegetation and are an important source of nutrients for ectomycorrhizal fungi [78]. Also, chitin is the main polysaccharide found in fungal cell walls and invertebrate exoskeletons [79] having significant quantities of nitrogen. These complex biopolymers are degraded enzymatically into simpler water-soluble forms of sugar through saprotrophic or mycorrhizal fungi reflecting their saprophytic ability which can be indirectly related to the survival of their host plant. For the survival of the host plant in wild, its nutritional demands for carbon and energy are met by the decomposition of this organic content present in the environment by OMF at the same site. Little information is available on the saprophytic behavior of OMF and more research is required to understand the nutritional physiology of both the partners by having a complete understanding of the role of OMF in decomposing the organic matter present in the ecosystem.

5.1.2 Litter degradation through enzymes

The decomposition of organic matter by saprotrophic basidiomycetes is a complex mechanism and does involve the participation of various enzymes and reactions. Saprophytic fungi stand apart from other organisms in their ability to decompose non-protein sources [73]. Various chlorophyllous and achlorophyllous orchid species are associated with saprophytic fungi from species of *Rhizoctonia* and *Epulorhiza* [80]. Utilization of these complex compounds in a litter is associated with the activity or production of extracellular enzymes (endo- or exo-) in basidiomyceteous fungi. These complex sources of carbon are degraded into their simpler forms through the activity of hydrolytic enzymes. Various litter components require a different set of enzymes for decomposition to occur such as cellobiohydrolases, Endo-1,4-β glucanases, and 1,4-β-glucosidases which effectively decompose cellulose to cellobiose. β-glucosidases then convert cellobiose to glucose.

Hemicelluloses are the second most abundant, heterogeneous polysaccharides present in the plant cell walls and comprise branched polymers of 500–3000 C5 or C6 sugars [81]. Lignin and plant cell wall polysaccharides (hemicellulose) interact with cellulose fibers to strengthen plant cell walls. Pectinases are widely produced by plant pathogens and endopolygalacturonase is one of the major enzymes involved in pathogenesis produced by a large number of pathogens such as *Rhiizoctonia solani* [82], *Phytophthora infestans* and *Verticillium* species [83]. Several pathogenic fungi degrade pectin and the release of these enzymes allows them to infect their host plant under favorable conditions but activates the cascade of defense reactions in plant cells [84].

Recent studies on OMF from Australian orchids, in the genera *Caladenia*, *Diuris*, *Drakaea* and *Pterostylis*, have shown utilization of pectin as a sole carbon source, resulting in the production of fungal biomass ranging from greater than to less than that on xylan [16]. Several extracellular enzymes, such as dehydrogenases and oxidases from the mycelium, are involved in wood-lignin decomposition and have the potential to utilize all major constituents of litter [81]. Microbial decomposition in heathland soils is a slow process [85] and the penetration of the resource is important [86]. Most wood-associated decay reactions occur close to fungal hyphae due to limited amounts of diffused enzymes [81] and lignocellulose-degrading units in the cell walls [87]. Burnett [88] proposed that enzyme secretion may occur in different areas of the apical region and these findings were further supported by experimental evidence in *Neurospora crassa,* where structural and physiological differences in the hyphal cell wall at the apical region contributed to the variation in secretion and retention of exoenzymes in the wall.

5.1.3 Breakdown of complex sources into soluble compounds and their use

In many ecosystems, most of the nutrients are locked up in organic compounds, soil microflora and microfauna. Organic macromolecules present in the soil are degraded to intermediate forms through the saprophytic ability of decomposers adding up to higher decay rates in the soil [89]. Some of the complex compounds in the form of cellulose, hemicelluloses (xylans and arabinoxylans), starch and pectin are degraded to soluble intermediate forms such as oligosaccharides, disaccharides, cellobiose, xylobiose and maltose which are finally broken down to their soluble breakdown products such as glucose, mannitol, trehalose, arabinose, galactose, mannose, xylose, rhamnose and glucuronic acid.

On penetrating a substrate, fungi decompose it and absorb its nutrients. The available nutrients help the fungus to grow and proliferate until the nutrients are depleted and fungus becomes dormant. In nature, succession starts at this point and other species feed on the remains. Succession in microorganisms is very important in completely digesting complex carbon sources to simple soluble compounds. The C:N ratio plays a vital role in determining microbial growth and the amount of decomposition taking place. Inter-relationships are sometimes antagonistic, with exploitation, antibiosis and competition being very common [89]. An average of 30–40% of C from decomposed substratum is assimilated by the fungi under favorable conditions [89].

OMF, as saprophytes, break down these complex macromolecules and transfer the intermediate and final soluble products to their hosts. The fungal partner increases the efficiency of the host plant in acquiring C, N and P from litter and soil.

So, it is important to understand the ability of OMF to utilize soluble carbon sources. Fungi break down complex molecules into intermediate and then simpler water-soluble forms. These soluble forms are then assimilated and used in metabolic pathways, or liberated as free metabolites, to be used by the OMF or competitive microorganisms, and may be subsequently transferred to the host plants.

5.1.3.1 Use of soluble carbon sources by OMF

Some of the soluble compounds released on the digestion of complex carbon sources are simpler soluble forms of sugars in the form of monosaccharides and disaccharides. OMF vary in their absorption of nutrients from the soil, similar to other mycorrhizal fungi. The ability of OMF to utilize a range of soluble carbon compounds has been studied previously but information available is fragmentary if compared to other mycorrhizal groups such as ERM and ECM fungi. Earlier physiological studies have stated that OMF metabolize sugars through an activity of enzymes such as and maltases and diastase-invertases [90]. There is little information available on the activity of enzymes and transporters involved in OM symbioses, but soluble carbon sources are likely to be transported rapidly to both pelotons and orchid cells which are later used in metabolism. Moreover, few studies have demonstrated the translocation and hydrolysis of the disaccharide sugar trehalose at the interface of the symbionts in MH orchids [91, 92]. Isotopic studies have shown a two-way transfer of carbon between the OMF and the orchid host [91, 93, 94] and it has recently been suggested that C and N containing compounds (derived from glucose and ammonium nitrate) are transferred from both senescent and live pelotons in *Spiranthes sinensis–Ceratobasidium* sp. AG-1 symbiosis *in vitro* [95].

To understand the potential of OMF to use soluble carbon sources requires their growth on a range of single carbon sources followed by measurement of their growth as fungal biomass. Research on Australian OMF has generally shown utilization of various soluble carbon sources such as the C5 arabinose, C6 glucose, C12 sucrose and cellobiose, and C(n) cellulose (as CMC), xylan, and pectin, and tannic acid [16–18]. Biomass on soluble carbon sources can be easily quantified by measuring the dry weight of mycelium and subtracting the biomass of controls from all the treatments, as used by Midgley et al. [16], Wright et al. [17] and Nurfadilah et al. [18] and Mehra et al. [75]. More recent studies have shown trends in utilization patterns of carbon sources across four fungal taxa from the *Rhizoctonia* alliance (*Ceratobasidium*, *Rhizoctonia*, *Tulasnella*, and *Serendipita*). OMF from these taxa produced large biomass on xylan, glucose, cellobiose, cellulose, pectin, and to some extent CMC, and the least fungal biomass was reported in all for tannic acid [18]. In studies on OMF from Australian orchids in the genera *Caladenia*, *Diuris*, *Drakaea* and *Pterostylis*, xylan consistently produced the greatest growth, often exceeding that on glucose [16–18].

For the establishment of balanced symbiosis between two partners more research using similar methods is required to determine the nutritional preferences displayed by OMF from other Australian terrestrial orchid species. The ability of an OMF to compete for and use soluble carbon compounds from sources external to the orchid may reflect the ability of its host orchid to survive and thrive.

5.2 Nitrogen sources

N present in the soil litter is typically found in the form of inorganic N (nitrates and ammonium) and organic N. Organic N comprises a large fraction of Australian litter but its utilization by OMF has been poorly studied. In the natural environment, amides and amino acids are easily accessible to the OMF, external to the orchid as a result of a litter breakdown and internally in the orchid as a result of plant metabolism.

The utilization of a wide range of organic and inorganic forms of nitrogen by OMF suggests their specificity of enzymes to hydrolyze complex forms of amides and peptides into simpler soluble organic N sources that are directly absorbed by OMF. The uptake and transfer of N by OMF has already been reported previously for northern hemisphere OMF [20, 96, 97] whereas, Cameron et al. [93] provided direct evidence of uptake and transfer of organic N through *Ceratobasidium cornigerum* (from *Goodyera repens*) by double-labeling of amino acid glycine. Recently, studies have also shown a transfer of N from the soil and through tripartite relationships by a single OMF of MH orchid, *Rhizanthella gardneri* [98]. Most recently, the uptake and transport of nitrogen from NH_4NO_3 was inferred from isotopic enrichment of ^{15}N in the pelotons and uninfected cells of *Spiranthes sinensis* protocorms using ultra-high spatial resolution secondary ion mass spectrometry (SIMS) [95]. With inorganic N sources, most authors reported greater utilization of NH_4^+ than NO_3^- in OMF strains of *Tulasnella* (one strain, *C. flava*) and *Serendipita* (six strains, *C. tentaculata*) whereas, many of these did not utilize nitrate [17, 18]. With organic sources, most OMF were capable of utilizing C3 alanine, C4 aspartic acid and/or asparagine, C5 glutamic acid and C6 arginine well as compared to C5 proline and C6 histidine which were poorly utilized [17, 18, 99]. Few OMF utilized C2 glycine well and others poorly; the latter included an isolate from *C. flava* [18]. In addition, only two out of six OMF from Australian *Pterostylis* species utilized tryptophan [16]. Recently, research work on Australian endangered orchid species (*C. fulva*) has shown that one of the symbiotically effective isolates, utilized most of the N sources with minimal variations in their biomass in contrary to the ineffective isolate under *in-vitro* conditions. The reason suggested for this was that it would affect their competition, at both levels in the host plant (internal/external) whereby, an ineffective isolate can successfully outcompete the effective isolate and its host, leading to chlorosis before the death of an earlier surviving orchid seedling [15].

5.3 Phosphorus sources

Most Australian soils are ancient and are phosphorus-deprived, as most of it has been leached out over time [100]. Along with N, it is one of the major limiting factors for plant growth. In soil, it is present in two major forms: inorganic P (P_i) in the form of phosphates where they are present in the form of scarcely available complexes [101] and mineral and organic phosphorus (P_o) as phosphate diesters, phosphate monoesters and inositol phosphates [100] where they are low in orthophosphate levels [102]. In natural environments, fungi degrade organic phosphorus compounds present in the dead matter but organic phosphorus locked in humus-rich forest soils is not easily accessible [100, 103]. Inorganic phosphorus has low solubility and is present in three main fractions: soil solution (dissolved phosphates), a labile pool (phosphates adsorbed to surfaces) and a non-labile pool (metal phosphates) [104].

Plants cannot utilize organic phosphates as they only have access to soluble phosphates and can readily absorb them [104]. Mycorrhizal associations can overcome nutrient limitations to plant growth by increasing the availability of phosphorus. Fungi can release phosphorus into the soil solution from organic phosphates with the help of phosphatases, thereby providing access for plants to otherwise insoluble forms of phosphorus [105]. The greater availability of phosphorus to the mycorrhizal plant host is dependent on the ability of its symbiont to absorb and translocate inorganic phosphates to the host roots and to access the forms of phosphorus 'locked up' in organic debris [106, 107]. Fungi can store phosphorus in their vacuoles as polyphosphate chains or as condensed phosphate [108].

Terrestrial orchid habitats are nutrient-deprived in Australia and leaf litter is among one of the major phosphorus sources available to OMF [100], through its richness in the cyclic phytic acid (inositol hexaphosphate, IP6, inositol polyphosphate), the main form of phosphorus storage in plants. In orchids it is assumed that mycorrhizal associations benefit the host plant by increasing the uptake of phosphorus. Earlier studies have reported the secretion of acid phosphatases by fungi in pure cultures [43]. The transfer of organic phosphorus in young protocorms of orchids through mycorrhizal fungi was first demonstrated by Smith [109] whereas, the uptake of inorganic phosphorus in mycorrhizal adult seedlings of *Goodyera repens* has been reported previously. Whilst, the utilization of organic phosphorus was demonstrated by Smith and Read [4] through the hydrolysis of organic compounds with a release of inorganic phosphorus (P_i). So far, there are few studies on the utilization of various forms of phosphorus by OMF in contrast to extensive work done on other mycorrhizae. A recent study by Nurfadilah et al. [18] showed that OMF from four genera of Australian orchids produced greater biomass with inorganic phosphate than DNA and with intermediate levels in case of phytic acid.

6. Ecological implications

Fungal preferences for specific carbon sources from the heterogeneous and unstable distribution of the substrates on forest floors might suggest that different stages of host plant development may have a preference for different organic substrates, for example, the abundance and presence of orchid seedlings (*Tipularia discolor*) near decaying logs in specified habitats as opposed to their absence near-adult flowering individuals [43] suggests that OMF does have preferences for their carbon sources, which could therefore explains their patchy distribution in the environment. The relative lack of utilization of some soluble components likely to be generated, may offer opportunities and niches for other fungi and microorganisms in general. OMF must compete not only with one another but also with other mycorrhizal and saprophyticic fungi for these resources, and for their breakdown products.

The relative abilities of OMF from Australian endangered and common orchid species (*Caladenia* spp.) to grow on the breakdown products of litter may have some ecological implications for their orchid hosts in terms of their taxonomy and conservation status [75]. Similarly, Nurfadilah et al. [18] concluded that the OMF from rare and common orchid species has the same utilization profiles of soluble carbon sources, having slow and uncompetitive growth could explain the conservation status of its host orchid. The importance of these nutritional studies can be related to the patchy spatial distribution of OMF and their host orchids [110]. Previous *in-vitro* studies showed competition between orchid siblings for available resources through their OMF and there is a possibility that this could be true for the orchids growing in the wild [111, 112].

Mehra et al. [75] showed that the OMF from various *Caladenia* species are differentiated not so much by different profiles of carbon sources utilized but by different rates of growth and final biomass. This suggests that threatened orchids contain OMF with relatively slow-growing and uncompetitive OMF compared with those from common orchids. It would be interesting to test this further by examining more OMF from a greater range of orchids. Also, *Ceratobasidium* species have rapid rates of growth compared with those of *Serendipita* and *Tulasnella*, the other two main OMF of Australian orchids, and it would be interesting to test these in direct competition in microcosms to see the effects on the survival of orchid seedlings of their respective hosts.

7. Conclusion

Orchids depend on their fungal endophytes for their nutritional demands, which is obligatory in its initial stages but may vary in adult green orchids, though they continue to harbor OMF in their underground organs. The forms of C, N, and P available to the OMF can determine their availability to the orchid host and can indirectly affect its conservation status. Thus, obtaining an effective symbiont is critical for an orchid's survival and is absolutely a high priority in recovery plans for endangered species. In order to develop effective strategies for conservation of orchids, a large number of orchid taxa should be tested for their nutritional modes as a function of their habitat based partly on organic content using labeling techniques and isotopic fractionations. Also, future research should be focused on developing enzymatic profiles for OMF using sterilized natural substrates and insoluble carbon sources, which may augment our understanding of the role of OMF in the decomposition of organic matter in the ecosystem. Uptake of soluble carbon sources in OMF from terrestrial green orchids can be further investigated through radiotracer techniques through labeling and setting up small microcosm experiments. Tracing the translocation of external highly enriched carbon sources over a short period of time will provide evidence on the net transfers of different forms of carbon between the OMF and the orchid.

Acknowledgements

Author acknowledges all the valuable guidance provided by Professor Ann C Lawrie and Dr Fiona Coates from RMIT University, Melbourne, Australia.

Author details

Shalika Mehra
RMIT University, Melbourne, Australia

*Address all correspondence to: shalika.mehra@rmit.edu.au

References

[1] IUCN. The IUCN Red List of Threatened Species: The IUCN Red List of Threatened Species; 2019. Available from: http://www.iucnredlist.org (updated 16 December)

[2] Chase MW, Cameron KM, Freudenstein JV, Pridgeon AM, Salazar G, Van den Berg C, et al. An updated classification of Orchidaceae. Botanical Journal of the Linnean Society. 2015;**177**(2):151-174

[3] Willis K. State of the World's Plants 2017. Report. Royal Botanic Gardens, Kew. 2017

[4] Smith SE, Read DJ. Mycorrhizal Symbiosis. 3rd ed. San Diego, CA: Academic Press; 2008

[5] Jones DL. A Complete Guide to Native Orchids of Australia, Including the Island Territories. Frenchs Forest, Australia: Reed New Holland; 2006

[6] Dixon K, Tremblay RL. Biology and natural history of *Caladenia*. Australian Journal of Botany. 2009;**57**:247-258

[7] Newman B, Ladd P, Batty A, Dixon K. Ecology of orchids in urban bushland reserves–can orchids be used as indicators of vegetation condition? Lankesteriana. 2015;**7**(1-2):313-315

[8] Gale SW, Fischer GA, Cribb PJ, Fay MF. Orchid Conservation: Bridging the Gap between Science and Practice. UK: Oxford University Press; 2018

[9] Fay MF. Orchid conservation: How can we meet the challenges in the twenty-first century? Botanical Studies. 2018;**59**(1):16

[10] Wraith J, Pickering C. A continental scale analysis of threats to orchids. Biological Conservation. 2019;**234**:7-17

[11] Jasinge N, Huynh T, Lawrie A. Changes in orchid populations and endophytic fungi with rainfall and prescribed burning in Pterostylis revoluta in Victoria, Australia. Annals of Botany. 2018;**121**(2):321-334

[12] Jasinge N, Huynh T, Lawrie A. Consequences of season of prescribed burning on two spring-flowering terrestrial orchids and their endophytic fungi. Australian Journal of Botany. 2018;**66**(4):298-312

[13] Lambers H, Brundrett MC, Raven JA, Hopper SD. Plant mineral nutrition in ancient landscapes: High plant species diversity on infertile soils is linked to functional diversity for nutritional strategies. Plant and Soil. 2010;**334**(1-2):11-31

[14] Hajong S, Kumaria S, Tandon P. Comparative study of key phosphorus and nitrogen metabolizing enzymes in mycorrhizal and non-mycorrhizal plants of *Dendrobium chrysanthum* Wall. ex Lindl. Acta Physiologiae Plantarum. 2013;**35**(7):2311-2322

[15] Mehra S. Nutritional and Genetic Diversity in Orchid Mycorrhizal Fungi from *Caladenia* Species. Melbourne: RMIT University; 2014

[16] Midgley DJ, Jordan LA, Saleeba JA, McGee PA. Utilisation of carbon substrates by orchid and ericoid mycorrhizal fungi from Australian dry sclerophyll forests. Mycorrhiza. 2006;**16**(3):175-182

[17] Wright M, Cross R, Cousens RD, May TW, McLean CB. The functional significance for the orchid *Caladenia tentaculata* of genetic and geographic variation in the mycorrhizal fungus *Sebacina vermifera* s. lat. complex. Muelleria. 2011;**29**(2):130-140

[18] Nurfadilah S, Swarts ND, Dixon KW, Lambers H, Merritt DJ. Variation in nutrient-acquisition patterns by

mycorrhizal fungi of rare and common orchids explains diversification in a global biodiversity hotspot. Annals of Botany. 2013;**111**(6):1233-1241

[19] Cameron DD, Johnson I, Read DJ, Leake JR. Giving and receiving: Measuring the carbon cost of mycorrhizas in the green orchid *Goodyera repens*. New Phytologist. 2008;**180**(1):176-184

[20] Gebauer G, Meyer M. ^{15}N and ^{13}C natural abundance of autotrophic and mycohetero-trophic orchids provides insight into nitrogen and carbon gain from fungal association. New Phytologist. 2003;**160**(1):209-223

[21] Julou T, Burghardt B, Gebauer G, Berveiller D, Damesin C, Selosse MA. Mixotrophy in orchids: Insights from a comparative study of green individuals and nonphotosynthetic individuals of *Cephalanthera damasonium*. New Phytologist. 2005;**166**(2):639-653

[22] Selosse MA, Roy M. Green plants that feed on fungi: Facts and questions about mixotrophy. Trends in Plant Science. 2009;**14**(2):64-70

[23] Merckx VSFT. Mycoheterotrophy: The Biology of Plants Living on Fungi. Springer; 2013. 356 p

[24] Pant B, Shah S, Shrestha R, Pandey S, Joshi PR. An overview on orchid endophytes. In: Mycorrhiza-Nutrient Uptake, Biocontrol, Ecorestoration. 4th ed. Cham, Switzerland: Springer; 2017. pp. 503-524

[25] Selosse MA, Bauer R, Moyersoen B. Basal hymenomycetes belonging to the Sebacinaceae are ectomycorrhizal on temperate deciduous trees. New Phytologist. 2002;**155**(1):183-195

[26] Dearnaley JDW, Martos F, Selosse MA. Orchid mycorrhizas: Molecular ecology, physiology, evolution and conservation aspects. In: Fungal Associations. The Mycota. Berlin Heidelberg: Springer; 2013. pp. 207-230

[27] Jacquemyn H, Duffy KJ, Selosse M-A. Biogeography of orchid mycorrhizas. Biogeography of Mycorrhizal Symbiosis. Springer; 2017;**230**:159-177

[28] Selosse MA. The latest news from biological interactions in orchids: in love, head to toe. New Phytologist. 2014;**202**(2):337-340

[29] Yagame T, Funabiki E, Nagasawa E, Fukiharu T, Iwase K. Identification and symbiotic ability of Psathyrellaceae fungi isolated from a photosynthetic orchid, *Cremastra appendiculata* (Orchidaceae). American Journal of Botany. 2013;**100**(9):1823-1830

[30] Weiß M, Waller F, Zuccaro A, Selosse MA. Sebacinales–one thousand and one interactions with land plants. New Phytologist. 2016;**211**(1):20-40

[31] Kohler A, Kuo A, Nagy LG, Morin E, Barry KW, Buscot F, et al. Convergent losses of decay mechanisms and rapid turnover of symbiosis genes in mycorrhizal mutualists. Nature Genetics. 2015;**47**(4):410-415

[32] Oliveira SF, Bocayuva MF, Veloso TGR, Bazzolli DMS, da Silva CC, Pereira OL, et al. Endophytic and mycorrhizal fungi associated with roots of endangered native orchids from the Atlantic Forest, Brazil. Mycorrhiza. 2014;**24**(1):55-64

[33] Allen TR, Millar T, Berch SM, Berbee ML. Culturing and direct DNA extraction find different fungi from the same ericoid mycorrhizal roots. New Phytologist. 2003;**160**:255-272

[34] Kottke I, Beiter A, Weiss M, Haug I, Oberwinkler F, Nebel M. Heterobasidiomycetes form symbiotic associations with hepatics: Jungermanniales have sebacinoid

mycobionts while *Aneura pinguis* (Metzgeriales) is associated with a*Tulasnella* species. Mycological Research. 2003;**107**(8):957-968

[35] Kottke I, Haug I, Setaro S, Suarez JP, Weiss M, Preuing M, et al. Guilds of mycorrhizal fungi and their relation to trees, ericads, orchids and liverworts in a neotropical mountain rain forest. Basic and Applied Ecology. 2008;**9**(1):13-23

[36] Weiss M, Sýkorová Z, Garnica S, Riess K, Martos F, Krause C, et al. Sebacinales everywhere: Previously overlooked ubiquitous fungal endophytes. PLoS One. 2011;**6**(2):e16793

[37] Selosse MA, Dubois MP, Alvarez N. Do Sebacinales commonly associate with plant roots as endophytes? Mycological Research. 2009;**113**:1062-1069

[38] Weiss M, Selosse M-A, Rexer K-H, Urban A, Oberwinkler F. Sebacinales: A hitherto overlooked cosm of heterobasidiomycetes with a broad mycorrhizal potential. Mycological Research. 2004;**108**(9):1003-1010

[39] Warcup JH. The mycorrhizal relationships of Australian orchids. New Phytologist. 1981;**87**(2):371-381

[40] Warcup JH, Talbot PHB. Perfect states of Rhizoctonia's associated with orchids. New Phytologist. 1967;**66**(4):631-641

[41] Warcup JH, Talbot PHB. Perfect states of Rhizoctonia's associated with orchids. II. New Phytologist. 1971;**70**(1):35-40

[42] Warcup JH. Mycorrhizal associations of isolates of *Sebacina vermifera*. New Phytologist. 1988;**110**(2):227-231

[43] Rasmussen HN. Terrestrial Orchids from Seed to Mycotrophic Plant. Cambridge, UK: Cambridge University Press; 1995

[44] Kristiansen KA, Taylor DL, Kjoller R, Rasmussen HN, Rosendahl S. Identification of mycorrhizal fungi from single pelotons of *Dactylorhiza majalis* (Orchidaceae) using single-strand conformation polymorphism and mitochondrial ribosomal large subunit DNA sequences. Molecular Ecology. 2001;**10**(8):2089-2093

[45] Wright MM, Cross R, Cousens RD, May TW, McLean CB. Taxonomic and functional characterisation of fungi from the *Sebacina vermifera* complex from common and rare orchids in the genus *Caladenia*. Mycorrhiza. 2010;**20**(6):375-390

[46] Warcup JH, Talbot PHB. Perfect states of some Rhizoctonia's. Transactions of the British Mycological Society. 1966;**49**:427-435

[47] Ramsay RR, Sivasithamparam K, Dixon KW. Anastomosis groups among *Rhizoctonia*-like endophytic fungi in southwestern Australian *Pterostylis* species (Orchidaceae). Lindleyana. 1987;**2**:161-166

[48] Andersen TF. A study of hyphal morphology in the form genus *Rhizoctonia*. Mycotaxon. 1990;**37**:25-46

[49] Warcup JH, Talbot PHB. Perfect states of Rhizoctonia's associated with orchids. III. New Phytologist. 1980;**86**(3):267-272

[50] Basiewicz M, Weiss M, Kogel K-H, Langen G, Zorn H, Zuccaro A. Molecular and phenotypic characterization of *Sebacina vermifera* strains associated with orchids, and the description of *Piriformospora williamsii* sp. nov. Fungal Biology. 2012;**116**(2):204-213

[51] Smith SE, Read DJ. Mycorrhizas in achlorophyllous plants (mycoheterotrophs). Mycorrhizal Symbiosis. Elsevier; 2008. pp. 458-506

[52] McKendrick SL, Leake JR, Read DJ. Symbiotic germination and development of myco-heterotrophic plants in nature: Transfer of carbon from ectomycorrhizal *Salix repens* and *Betula pendula* to the orchid *Corallorhiza trifida* through shared hyphal connections. New Phytologist. 2000;**145**(3):539-548

[53] Selosse MA, WE M, Jany JL, Tillier A. Communities and populations of sebacinoid basidiomycetes associated with the achlorophyllous orchid *Neottia nidus-avis* (L.) L.C.M. rich. And neighbouring tree ectomycorrhizae. Molecular Ecology. 2002;**11**(9):1831-1844

[54] Cameron DD, Preiss K, Gebauer G, Read DJ. The chlorophyll-containing orchid *Corallorhiza trifida* derives little carbon through photosynthesis. New Phytologist. 2009;**183**(2):358-364

[55] Ogura-Tsujita Y, Gebauer G, Hashimoto T, Umata H, Yukawa T. Evidence for novel and specialized mycorrhizal parasitism: The orchid *Gastrodia confusa* gains carbon from saprotrophic *Mycena*. Proceedings of the Royal Society B: Biological Sciences. 2009;**276**(1657):761-767

[56] Xu JT, Mu C. The relation between growth of *Gastrodia elata* protocorms and fungi. Acta Botanica Singapore. 1990;**32**:26-31

[57] Yamato M, Iwase K, Yagame T, Suzuki A. Isolation and identification of mycorrhizal fungi associating with an achlorophyllous plant, *Epipogium roseum* (Orchidaceae). Mycoscience. 2005;**46**(2):73-77

[58] Ogura-Tsujita Y, Yukawa T. *In situ* seed sowing techniques for the recovery of endangered orchids. Japanese Journal of Conservation Ecology. 2008;**13**(1):121-127

[59] Shefferson RP, Weiss M, Kull T, Taylor DL. High specificity generally characterizes mycorrhizal association in rare lady's slipper orchids, genus *Cypripedium*. Molecular Ecology. 2005;**14**(2):613-626

[60] Taylor DL, Bruns TD, Szaro TM, Hodges SA. Divergence in mycorrhizal specialization within *Hexalectris spicata* (Orchidaceae), a nonphotosynthetic desert orchid. American Journal of Botany. 2003;**90**(8):1168-1179

[61] McKendrick SL, Leake JR, Taylor DL, Read DJ. Symbiotic germination and development of the myco-heterotrophic orchid *Neottia nidus-avis* in nature and its requirement for locally distributed *Sebacina* spp. New Phytologist. 2002;**154**:233-247

[62] Masuhara G, Katsuya K. *In situ* and *in vitro* specificity between *Rhizoctonia* spp. and *Spiranthes sinensis* (Persoon) Ames. var. *amoena* (M. Bieberstein) Hara (Orchidaceae). New Phytologist. 1994;**127**:711-718

[63] McCormick MK, Whigham DF, O'Neill J. Mycorrhizal diversity in photosynthetic terrestrial orchids. New Phytologist. 2004;**163**(2):425-438

[64] Dearnaley JDW. Further advances in orchid mycorrhizal research. Mycorrhiza. 2007;**17**(6):475-486

[65] Huynh TT, McLean CB, Coates F, Lawrie AC. Effect of developmental stage and peloton morphology on success in isolation of mycorrhizal fungi in *Caladenia formosa* (Orchidaceae). Australian Journal of Botany. 2004;**52**(2):231-241

[66] Bonnardeaux Y, Brundrett M, Batty A, Dixon K, Koch J, Sivasithamparam K. Diversity of mycorrhizal fungi of terrestrial orchids: Compatibility webs, brief encounters, lasting relationships and alien invasions. Mycological Research. 2007;**111**:51-61

[67] Peterson LR, Uetake Y, Zelmer C. Fungal symbioses with orchid protocorms. Symbiosis. 1998;**25**:29-55

[68] Feuerherdt L, Petit S, Jusaitis M. Distribution of mycorrhizal fungus associated with the endangered pink-lipped spider orchid *Arachnorchis* (syn. *Caladenia behrii*) at Warren Conservation Park in South Australia. New Zealand Journal of Botany. 2005;**43**(2):367-371

[69] Huynh TT, Thomson R, McLean CB, Lawrie AC. Functional and genetic diversity of mycorrhizal fungi from single plants of *Caladenia formosa* (Orchidaceae). Annals of Botany. 2009;**104**(4):757-765

[70] McCormick MK, Whigham DF, Sloan D, O'Malley K, Hodkinson B. Orchid-fungus fidelity: A marriage meant to last? Ecology. 2006;**87**(4):903-911

[71] Fricker MD, Bebber D, Boddy L. Mycelial networks: Structure and dynamics. British Mycological Society Symposia Series. Elsevier; 2008;**28**: 3-18

[72] Finlay R. Action and interaction in the mycorrhizal hyphosphere—A re-evaluation of the role of mycorrhizas in nutrient acquisition and plant ecology. In: BassiriRad H, editor. Nutrient Acquisition by Plants, Ecological Studies. Vol. 181. Berlin Heidelberg: Springer; 2005. pp. 221-276

[73] Boddy L, Frankland J, van West P. Ecology of Saprotrophic Basidiomycetes. Elsevier; 2008. 386 p

[74] Jakobsen I. Transport of phosphorus and carbon in arbuscular mycorrhizas. In: Varma A, Hock B, editors. Mycorrhiza: Structure, Function, Molecular Biology. 2nd ed. Heidelberg: Springer; 1999. pp. 535-542

[75] Mehra S, Morrison P, Coates F, Lawrie A. Differences in carbon source utilisation by orchid mycorrhizal fungi from common and endangered species of *Caladenia* (Orchidaceae). Mycorrhiza. 2017;**27**(2):95-108

[76] McDonald N. Growth of Mycorrhizal Fungi and Effects on their Plant and Environment [Honours]. Melbourne: RMIT; 2000

[77] De Vries RP, Visser J. *Aspergillus* enzymes involved in degradation of plant cell wall polysaccharides. Microbiology and Molecular Biology Reviews. 2001;**65**(4):497-522

[78] Perotto S, Coisson JD, Perugini I, Cometti V, Bonfante P. Production of pectin-degrading enzymes by ericoid mycorrhizal fungi. New Phytologist. 1997;**135**(1):151-162

[79] Mitchell DT, Sweeney M, Kennedy A. Chitin degradation by *Hymenoscyphus ericae* and the influence of *H. ericae* on the growth of ectomycorrhizal fungi. In: Read DJ, Lewis DH, Fitter AH, Alexander IJ, editors. Mycorrhizas in Ecosystems. Wallingford, UK: CAB International; 1992. pp. 246-251

[80] Currah RS, Smreciu EA, Hambleton S. Mycorrhizae and mycorrhizal fungi of boreal species of *Platanthera* and *Coeloglossum* (Orchidaceae). Canadian Journal of Botany. 1990;**68**:1171-1181

[81] Baldrian P. Enzymes of saprotrophic basidiomycetes. In: Ecology of Saprotrophic Basidiomycetes. Elsevier; 2008. pp. 19-41

[82] Ayers WA, Papaviza GC, Diem AF. Polygalacturonate trans-eliminase and polygalacturonase production by *Rhizoctonia solani*. Phytopathology. 1966;**56**(9):1006

[83] Ward OP, Moo-Young M, Venkat K. Enzymatic degradation of cell wall and related plant polysaccharides. Critical Reviews in Biotechnology. 1989;**8**(4):237-274

[84] Hahn MG, Darvill AG, Albersheim P. Host-pathogen

interactions XIX. The endogenous elicitor, a fragment of a plant cell wall polysaccharide that elicits phytoalexin accumulation in soybeans. Plant Physiology. 1981;**68**(5):1161-1169

[85] Read DJ, Mitchell DT. Decomposition and mineralization processes in mediterranean-type ecosystems and in heathlands of similar structure. In: Mediterranean-Type Ecosystems: The Role of Nutrients. Springer-Verlag; 1983. pp. 208-232

[86] Boddy L. Interspecific combative interactions between wood-decaying basidiomycetes. FEMS Microbiology Ecology. 2000;**31**(3):185-194

[87] Valášková V, Baldrian P. Estimation of bound and free fractions of lignocellulose-degrading enzymes of wood-rotting fungi *Pleurotus ostreatus*,*Trametes versicolor* and *Piptoporus betulinu*. Research in Microbiology. 2006;**157**(2):119-124

[88] Burnett JH, editor. Fundamentals of Mycology. New York: St. Martin's Press; 1968

[89] Moore-Landecker E. In: Carl PS, editor. Fundamentals of the Fungi. Englewood Cliffs, NJ, USA: Prentice-Hall, Inc.; 1972. p. 482

[90] Burgeff H. Die Wurzelpilze der Orchideen, ihre Kultur und ihre Leben in der Pflanze. Jena: Gustav Fischer; 1909

[91] Smith SE. Carbohydrate translocation in orchid mycorrhizas. New Phytologist. 1967;**66**(3):371-378

[92] Smith SE, Smith FA. Structure and function of the interfaces in biotrophic symbioses as they relate to nutrient transport. New Phytologist. 1990;**114**(1):1-38

[93] Cameron DD, Leake JR, Read DJ. Mutualistic mycorrhiza in orchids: Evidence from plant-fungus carbon and nitrogen transfers in the green-leaved terrestrial orchid *Goodyera repens*. New Phytologist. 2006;**171**(2):405-416

[94] Cameron DD, Johnson I, Leake JR, Read DJ. Mycorrhizal acquisition of inorganic phosphorus by the green-leaved terrestrial orchid *Goodyera repens*. Annals of Botany. 2007;**99**(5):831-834

[95] Kuga Y, Sakamoto N, Yurimoto H. Stable isotope cellular imaging reveals that both live and degenerating fungal pelotons transfer carbon and nitrogen to orchid protocorms. New Phytologist. 2014;**202**(2):594-605

[96] Bidartondo MI, Burghardt B, Gebauer G, Bruns TD, Read DJ. Changing partners in the dark: Isotopic and molecular evidence of ectomycorrhizal liaisons between forest orchids and trees. Proceedings of the Royal Society of London. Series B: Biological Sciences. 2004;**271**(1550):1799-1806

[97] Leake JR. Myco-heterotroph/epiparasitic plant interactions with ectomycorrhizal and arbuscular mycorrhizal fungi. Current Opinion in Plant Biology. 2004;**7**(4):422-428

[98] Bougoure JJ, Brundrett MC, Grierson PF. Carbon and nitrogen supply to the underground orchid Rhizanthella gardneri. New Phytologist. 2010;**186**(4):947-956

[99] Stephen RC, Fung KK. Nitrogen requirements of the fungal endophytes of *Arundina chinensis*. Canadian Journal of Botany. 1971;**49**(3):407-410

[100] Lambers H, Raven JA, Shaver GR, Smith SE. Plant nutrient-acquisition strategies change with soil age. Trends in Ecology & Evolution. 2008;**23**(2):95-103

[101] Shen J, Yuan L, Zhang J, Li H, Bai Z, Chen X, et al. Phosphorus

dynamics: From soil to plant. Plant Physiology. 2011;**156**(3):997-1005

[102] Attiwill PM, Leeper GW. Forest Soils and Nutrient Cycles. Melbourne: Melbourne University press; 1987

[103] Cairney J. Ectomycorrhizal fungi: The symbiotic route to the root for phosphorus in forest soils. Plant and Soil. 2011;**344**(1):51-71

[104] Jennings DH, Lysek G. Fungal Biology: Understanding the Fungal Lifestyle. Oxford, UK: BIOS Scientific Publishers Ltd; 1996

[105] Joner EJ, Johansen A. Phosphatase activity of external hyphae of two arbuscular mycorrhizal fungi. Mycological Research. 2000;**104**(01):81-86

[106] Finlay RD, Read DJ. The structure and function of the vegetative mycelium of ectomycorrhizal plants. II. The uptake and distribution of phosphorus by mycelium inter-connecting host plants. New Phytologist. 1986;**103**(1):157-165

[107] Sawyer NA, Chambers SM, Cairney JWG. Utilisation of inorganic and organic phosphorus sources by isolates of *Amanita muscaria* and *Amanita* species native to temperate eastern Australia. Australian Journal of Botany. 2003;**51**(2):151-158

[108] Bolan NS. A critical review on the role of *mycorrhizal* fungi in the uptake of phosphorus by plants. Plant and Soil. 1991;**134**(2):189-207

[109] Smith SE. Physiology and ecology of orchid *mycorrhizal* fungi with reference to seedling nutrition. New Phytologist. 1966;**65**:488-499

[110] Wardle DA. A comparative assessment of factors which influence microbial biomass carbon and nitrogen levels in soil. Biological Reviews. 1992;**67**(3):321-358

[111] Batty AL, Dixon KW, Brundrett MC, Sivasithamparam K. Orchid conservation and mycorrhizal associations. In: Microrganisms in Plant Conservation and Biodiversity. The Netherlands: Springer; 2002. pp. 195-226

[112] Rasmussen H, Johansen B. Density-dependent interactions between seedlings of *Dactylorhiza majalis* (Orchidaceae) in symbiotic *in vitro* culture. Physiologia Plantarum. 1989;77(4):473-478

Chapter 2

Impact Brassinolide on Two Fig Varieties

Zulias Mardinata, Mardaleni and Tengku Edy Sabli

Abstract

Brassinolide (BL) is a plant hormone showing wide occurrence in the plant kingdom with unique biological effects on growth and physiological traits. The fig varieties, Improved Brown Turkey (IBT) and Masui Dauphine (MD), are commonly found in Indonesia and Malaysia. There is limited information on exogenous brassinolide application on these varieties. In this chapter, we present the effect of different concentration of exogenous application of BL on growth and physiological changes of fig. Increasing BL concentration (50, 100, and 200 ml.L^{-1}) caused some differences in growth and physiological changes of fig, but the differences were not consistent and most of the changes happened only in first or second month. Cultivar IBT showed higher growth and physiological changes than cultivar MD after receiving brassinolide treatment. There was significant effect of interaction between brassinolide and variety on growth and physiological changes of fig except in plant height and total dry biomass.

Keywords: Brassinolide, fig, growth, physiological changes

1. Introduction

Brassinolide (BL) is one of the brassinosteroids, which are steroidal plant hormones showing a wide occurrence in the plant kingdom, that have unique biological effects on growth and development [1, 2]. They are a group of naturally occurring polyhydroxy steroids initially isolated from *Brassica napus* pollen in 1979. Research on brassinosteroids has revealed that they elicit a wide spectrum of morphological and physiological responses in plants that include stem elongation and cell division [3], leaf bending and epinasty [4]. Besides their role in promoting plant growth activities, they also have physiological effects on the growth and development of plants [2, 5].

Much has been written about Clouse [6], for example, pointed out that:

Among plant hormones, BL are structurally the most similar to animal steroid hormones, which have well-known functions in regulating embryonic and post-embryonic development and adult homeostasis. Like their animal counterparts, BL regulate the expression of numerous genes, impact the activity of complex metabolic pathways, contribute to the regulation of cell division and differentiation, and help control overall developmental programs leading to morphogenesis. They are also involved in regulating processes more specific to plant growth including flowering and cell expansion in the presence of a potentially growth-limiting cell wall (p. 1).

Fig (*Ficus carica* L.) belongs to the *Moraceae* family. It is a bush or small tree, moderate in size, deciduous with broad, ovate, three- to five-lobed leaves, contains copious milky latex and introduced to Indonesia and Malaysia from Middle East and Western Asia. There are over 700 named varieties of fig trees, but many of them are not grown in home garden [7]. Because fig seeds are non-viable, trees must be propagated via cuttings or grafts. Though the propagation of *F. carica* by vegetative cuttings insures uniformity, relatively low multiplication rates are achieved because these materials can be obtained only from upright branches, which results in poor rooting [8]; hence, brassinolide application was attempted by evaluating plant growth and physiological changes in *Ficus carica*.

In Malaysia and Indonesia, there are at least 21 known varieties of the fig tree and most of them are from Improved Brown Turkey (IBT) and Masui Dauphine (MD) varieties [9]. There is limited information on exogenous brassinolide application on these varieties.

2. Brassinolide

Brassinolide (BL) or 2,3,22,23-Tetrahydroxy-β-homo-7-oxaergostan-6-one or $C_{28}H_{48}O_6$ with molar mass 480.69 g mol^{-1} is a plant hormone [10]. The first isolated brassinosteroid (BRs), it was discovered when it was shown that pollen from rapeseed (*Brassica napus*) could promote stem elongation and cell division. The biologically active component was isolated and named BL [3].

Common structural characteristics of BL (**Figure 1**) are A/B *cis* combined steroidal skeletons, oxygenated task in rings A, the lateral chain and with very few exceptions the ring B. All the naturally happening BRs apply the same qualitative results as BL. The biological activity magnitude has been found to rely on the location and spatial orientation of the hydroxyl groups in ring A, the oxygenated function nature locate at ring B, the configuration of chiral centers C_{22} and C_{23} and the substitution pattern and configuration of the chiral center C_{24}. Attempts to expand a Quantitative Structure–Activity Relationship (QSAR) system that enable to look the biological action of a given compound have been developed [11].

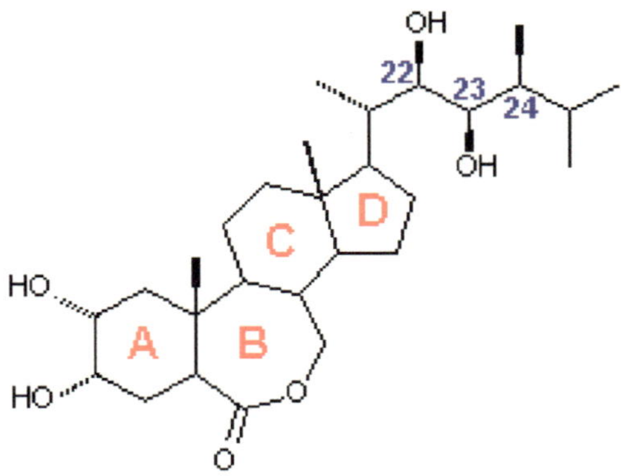

Figure 1.
Chemical structure of brassinolide [10].

3. History of Brassinolide

BL for the first time was found in 1968. To produce a strong plant growth promoting effect, Marumo et al. [12] gained three extracts from an evergreen Japanese plant known as Isonuki (*Distylium rasemosun* Sieb et Zucc.). A couple years later, Mitchell et al. [10] used rape pollen *Brassica napus* L. resulting an oil fraction to use the same effect. They guessed that the compounds that yielded such effects might be a plant hormones new kind, but at that time no structural characterization was possible caused by the limited amount of material availability.

Grove et al. [3] purified 40 kg of collected bee pollen of *Brassica napus* L. to establish its structure, which shows resemblance to animal steroid hormones resulted 4 mg of a crystalline ingredient which was known as (22R,23R,24S)-2a,3a,22,23-tetrahydroxy- 24-methyl-B-homo-7-oxa-5a-cholestan-6-one. In consequence of its low concentration, the BL identification took 10 years of loyal work at U.S. Department of Agriculture at a cost of over 1 million U.S. dollars [13]. This steroidal lactone which stimulates plant development when utilized to plants in concentrations lower than 10^{-4} mg ml^{-1} received the name of BL, product of the combination of Brassica, the Latin name of the plant from which it was isolated and the suffix '*-olide*' characteristic of the lactones. After isolation and identification of BL a number of manufacturally connected steroids have been isolated [14]. Till 2001 year, over 40 naturally happening BRs are identified; these bring an oxygen moiety at the C-3 location in fusion with others at the C-2, C-6, C-22 or C-23 locations [15].

4. Roles of Brassinolide in plant

According to Tang et al. [16], benefits of BL improves the growth of the germinating seed, improves the plant's ability to deal with stress, such as diseases, drought, salt, and cold, promotes growth of lateral buds, produces deep green leaves, increases the number of flowers and fruit produced, increases the percentage of fruit setting by decreasing the amount of flower and fruit drop, increases sugar content and generally improves the quality of the fruit, promotes fruit enlargement, delays leaf senescence lengthening productivity, and can be used in soil, hydroponics or leaf feeding.

5. *Ficus carica* L.

The genus *Ficus* (**Table 1**) is remarkable for the variation among its 13 species, and is widely distributed throughout the tropics and subtropics of both hemispheres. Several species produce edible figs of varying palatability, whereas many other arborescent *Ficus* sp. are cultivated for shade, as avenue trees and as ornamentals. *Ficus carica* L. (**Figure 2**) is considered to be native to South-Western Asia and to have been brought into cultivation in the Southern Arabian Peninsula by 3000 BC [17]. The fig tree is small or moderate in size, and deciduous with broad, ovate, three- to five-lobed leaves. The leaves are rough above and pubescent below, long stalked and dark green with pronounced venation. The female fig plants have larger leaves and more spreading crowns than male trees [18].

The fig is an accumulation fruit structured by individual small drupes; known as a drupelet. In ovaries, the drupelets develop to a syconium containing amounts of

Rank	Scientific name
Kingdom	Plantae
Subkingdom	Tracheobionta
Division	Magnoliophyta
Class	Magnoliopsida
Subclass	Alismidae
Order	Rosales
Family	Moraceae
Genus	*Ficus*
Species	*Ficus carica*
Generic Group	*Mulberry*

Source: [21].

Table 1.
Taxonomy of Ficus carica.

Figure 2.
Part of Ficus carica *L.: (a) five-lobed leaf; (b) fig fruit; (c) female flower; (d) male flower; and (e) inside view of fig fruit [19].*

unisexual flowers and pollinated by wasps via ostiole. The fig may produce multiple crops of annual fruits and need pollen from their pollinator Capri figs in any certain fig types. Breba crop is borne laterally on the growth of the previous season from buds produced in leaf axils and it is not produced in all cultivars. Another fig fruit type is main crop which is produced laterally in the axils of leaves on current season shoots. The leaves fall and the tree enters the dormancy period at the end of the growth period [20].

Figs respond well to heavy mulching with organic materials to conserve moisture, improve soil structure and reduce root knot nematode levels but intolerant in condition of poorly drained and waterlogged. To increase the main crop and maintain size control, fig also responds well to pruning and can be trained or pruned heavily in the dormant season [21]. The fig is fairly salt and drought tolerant. Soils of high lime content produce fruits of better quality suitable for drying. Fig trees can also withstand temperatures as low as −12 to −9.5°C [18].

6. Cultivar improved Brown Turkey (IBT) and Masui dauphine (MD)

In Malaysia and Indonesia, there are at least of 21 known fig varieties being found [9] over 700 named varieties around the world, but many of them are of no

(a) (b)

Figure 3.
Cultivars of Ficus carica *L.: (a) Improved Brown Turkey (IBT); (b) Masui Dauphine (MD).*

Item	Improved Brown Turkey (IBT)	Masui Dauphine (MD)
Origin	Turkey	Japan
Skin color	Brownish-purple	Reddish-brown
Flesh color	Pink	Red strawberry
Tree height	4.5–7.5 m	2.5–3 m
Tree width	3.6–4.5 m	3–4 m
Drought tolerance	Good	Good
Annual pruning	Light	Light
Breba crop	Yes	Yes
Fruit weight	50–60 g	110–220 g
Harvesting	Twice/year	Twice/year
Leaf type	Five lobe	Three to five lobe
Pollination	Self-pollinating (common fig)	Self-pollinating (common fig)

Source: [22].

Table 2.
Comparison of two cultivar of Ficus carica.

use to home gardeners [7]. In this study, researcher use two varieties of common fig, they are Improved Brown Turkey (IBT) and Masui Dauphine (MD). Each variety has different key characteristics as shown in **Figure 3** and **Table 2**.

Most people choose to grow fruit trees in containers for easy mobility. Another reason is easy to measure plant quality. Plant quality is divided into three broad categories of attributes including morphological, physiological, and performance.

Morphological attributes are easy to see and measure and does not change readily after plants are harvested and stored [23]. Figs are well-suited to container cultivation. For this purpose, the ideal container size is about 10–15 gallons-substantial enough to support a tree, but small enough to move easily. Container grown figs need regular watering and feeding. Figs will not grow for very long if it does not have adequate drainage. Make sure the container that we use has holes (usually in the bottom and/or sides), so that any excess water can drain, and air can access the soil. This will help us to prevent potentially fatal diseases like root rot [24].

7. Effect Brassinolide on growth of fig

The growth of the fig plants was affected by the brassinolide levels. Treatment of the fig plants with different concentrations of brassinolide (50, 100 and 200 ml.L^{-1}) caused an increase in plant height and total dry biomass compared to control samples. Total leaf area, specific leaf area and shoot-to-root ratio increased with increasing concentrations of brassinolide up to 100 ml.L^{-1}, followed by a decline whereas net assimilation rate fluctuated over a period of study. At the first Month After Treatment (MAT), increasing brassinolide concentration (50 and 100 ml.L^{-1}) caused an increase in the net assimilation rate when compared to control but there was a decrease when brassinolide concentration was 200 ml.L^{-1}. At the second MAT, by increasing the brassinolide concentration (50, 100 and 200 ml.L^{-1}), had decreased the net assimilation rate.

Application of brassinolide had some effect on plant height, total leaf area, total dry biomass, specific leaf area and net assimilation rate but it was not significant on the shoot-to-root ratio. Among the varieties, IBT showed higher growth than MD at every five-weekly observation. There was a significant interaction between the brassinolide and the cultivar for total leaf area, specific leaf area, shoot-to-root ratio and net assimilation rate parameters. Additionally, only shoot-to-root ratio parameter showed a significant effect of interaction between the brassinolide and cultivar at 1% level of significance.

The effect of exogenous brassinolide application on some growth and physicological traits on two cultivars of fig has been investigated. The main functions of brassinolide are to promote the plant growth especially for cell elongation and division [25] and has the ability to stimulate other physiological processes [26]. Wang et al. [27] had found that brassinollide appeared to cause elongation by affecting wall extensibility and increasing wall relaxation properties.

As levels of brassinolide increased (50, 100 and 200 ml.L^{-1}), plant height, leaf area, total dry biomass and net assimilation rate parameters also linearly improved at 28, 25, 6 and 66%, respectively, higher than recorded for the control treatment. Similar results were reported by other researchers for other plants, i.e., Hu et al. [28] for *Leymus chinensis*; Bera et al. [29] for sunflower; and Anjum et al. [30] for maize. The growth stimulation was more pronounced on above ground biomass than below ground biomass, showing a high shoot-to-root ratio [31]. The increase in growth in this study might have been due to increased carboxylation rate after using the BL treatment, which enhanced carbon assimilation, channeling it to stimulate increase in plant height, leaf area and total biomass [32].

Specific leaf area (SLA) is one growth parameter that characterized the thickness of the leaves. Usually plant with high SLA had the thinnest leaves. Specific leaf area was found to be lower than the control ($p \leq 0.05$) under brassinolide concentrations of 50 and 100 $ml.L^{-1}$. The result implies that plants have thicker leaves. The thicker leaf might have been due to increase in the mesophyll layer after receiving brassinolide [33]. The increase in leaf thickness could also have been due to higher leaf weight ratio in fourth MAT compared with first to third MAT. The leaf area was maintained at lowest SLA. That indicated that leaves of fig were thickest at brassinolide 100 $ml.L^{-1}$. This indicated that increase in SLA was due to increase in leaf weight compared with increase in leaf area [34, 35].

The net assimilation rate (NAR) of plants are growth characteristics that best describe plant growth performance under specified conditions [36]. It is evident that plants under elevated BL have high NAR. Increase in plant growth grown under different planting geometries and depths in SRI has also been reported by Rajput et al. [37], who reported that increase in total biomass by 30% in rice had increased NAR by 4% compared with the control. The reduction in NAR was due to the ontogenical development of fig.

8. Effect of Brassinolide on physiological changes of fig

The physiological changes of fig were affected by the brassinolide levels and the cultivars. Interaction between brassinolide concentrations and fig variety was significant only at 5%. As like morphological parameters, physiological traits such as photosynthesis, transpiration rate, and chlorophyll have shown some differences with brassinolide application, but the differences were not consistent and most of the changes happened only in first or second month. Both the brassinolide and the cultivar treatments were effective on the physiological changes of fig except on stomatal conductance.

Varietal performance of brassinolide application was analyzed at specific period of the study and the result is presented in **Figures 4** and **5**. Increasing concentration of brassinolide (50, 100 and 200 $ml.L^{-1}$) had decreased the rate of photosynthesis, transpiration and chlorophyll content in IBT than MD.

BL had profound impact on leaf photosynthesis and plant performance. BL improved leaf carbon assimilation rate, which is the light harvesting machine of plant photosynthesis. BL treatment also enhanced photosynthetic performance of cotton seedlings under NaCl stress [38–40]. For cucumber seedlings, BL treatment has also been found to promote the occurrence of new roots, the formation of lateral roots and nutrient uptake [41].

BL-treatment enhanced photosynthesis (17.06%) and chlorophyll content (18.36%). In contrast, BL-treatment decreased stomatal conductance (11.94%) and transpiration rate (17.83%). The BL-induced increase in photosynthesis could have been due to improvements in leaf-water balance as indicated by increased water potential [42] and improved chlorophyll content and higher leaf area in BL-treated plants [43].

Stomata are the windows that admit water and CO_2 in and out of the plant. Chlorophyll content and transpiration rate were found to have declined. This could be attributed to the enhanced growth of seedlings under elevated BL treatment that diluted the nitrogen content in the plant tissue [44]. **Figure 6A** and **C** showed a significant positive inter-relation among chlorophyll content, transpiration

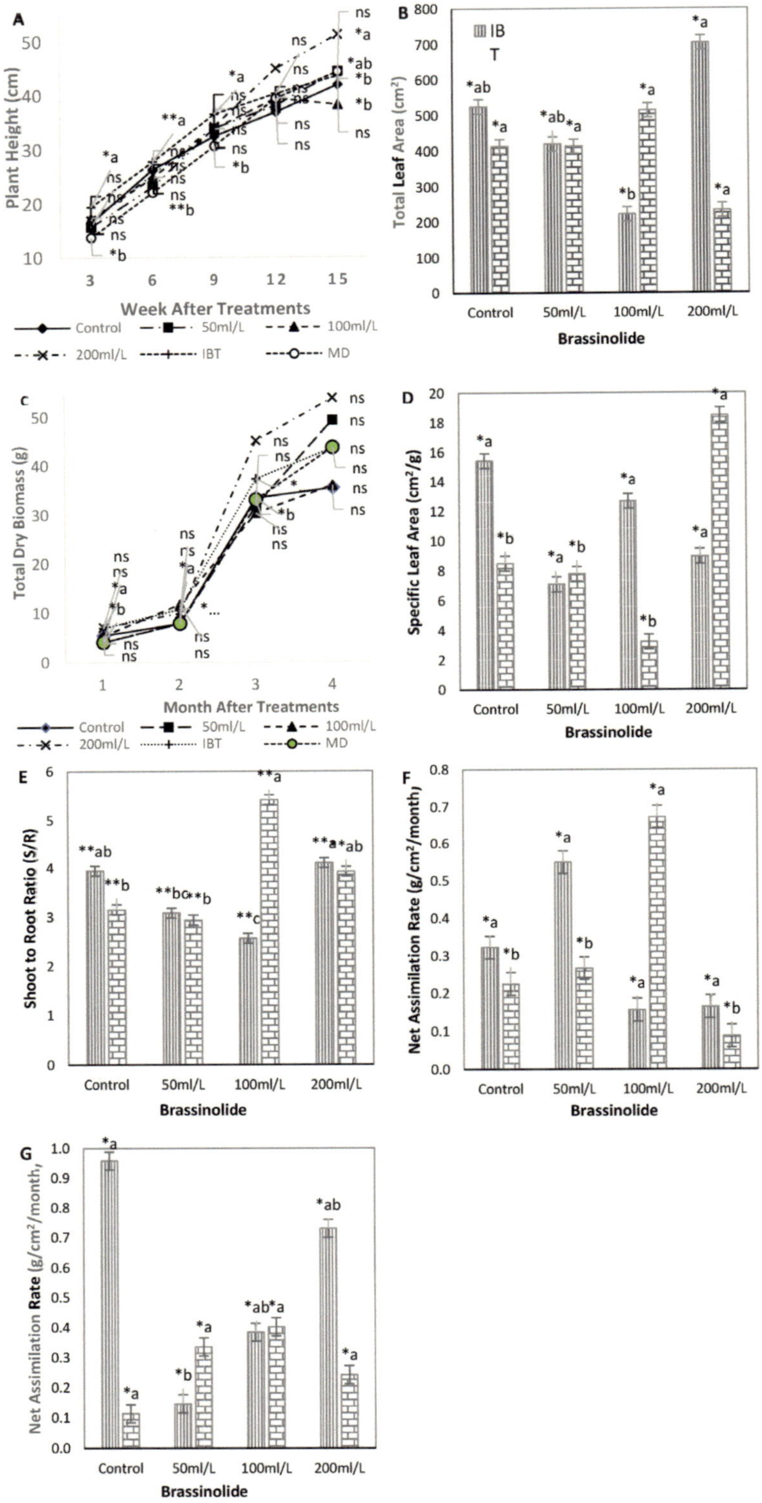

Figure 4.

Significant growth of fig according to parameters: (A) plant height as main effect of brassinolides on the cultivars; (B) TLA at third MAT as interaction between cultivars and brassinolide; (C) TDB as main effect of brassinolides and cultivars; (D) SLA at first MAT as interaction between cultivars and brassinolides; (E) S/R at fourth MAT as interaction between cultivars and brassinolides; NAR as interaction between cultivars and brassinolides at: (F) first MAT; and (G) second MAT. Bars and curves represent means followed by the different small letters are significant at * = $p < 0.05$, ** = $p < 1\%$, *and ns = not significant.*

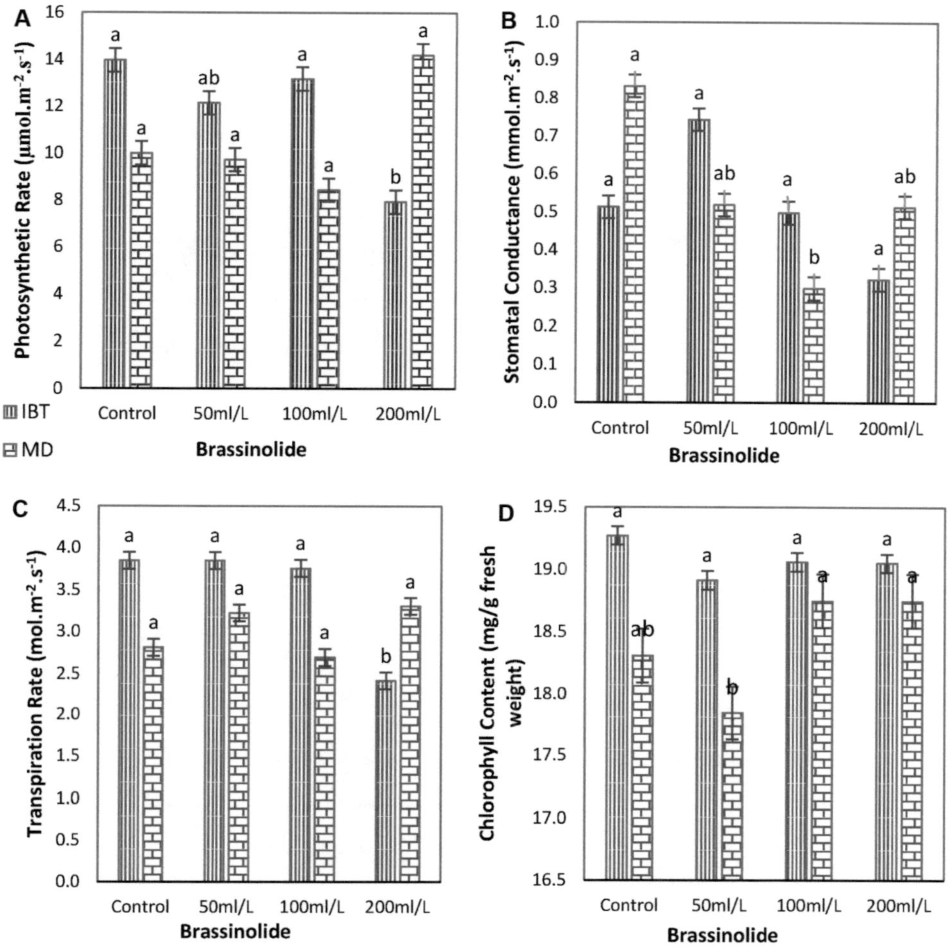

Figure 5.
Significant physiological changes of fig according to parameters: (A) A at second MAT; (B) gs at first MAT; (C) E at second MAT; and (D) CC at first MAT. Bars represent means followed by the different small letters significant at $p < 0.05$.

rate and stomatal conductance, indicating that a decrease in chlorophyll content would associated with same degree of reduction in transpiration rate and stomatal conductance.

9. Correlation analysis

Correlation analysis was carried out to establish the relationship between the parameters. **Figure 6** shows that a significant positive inter-correlation among parameters such as chlorophyll content, specific leaf area, transpiration rate and stomatal conductance. Increase in chlorophyll content, transpiration rate, total dry biomass, photosynthetic rate, and total dry biomass was associated with an increase in specific leaf area, transpiration rate, stomatal conductance, net assimilation rate and total leaf area with an r value of 14.95, 27.75, 3.97, 62.08, 36.93, 25.27 and 21.13%, respectively.

Significant negative correlation was noted between total dry biomass with specific leaf area; total dry biomass with transpiration rate; transpiration rate with net assimilation rate; chlorophyll content with net assimilation rate; and specific

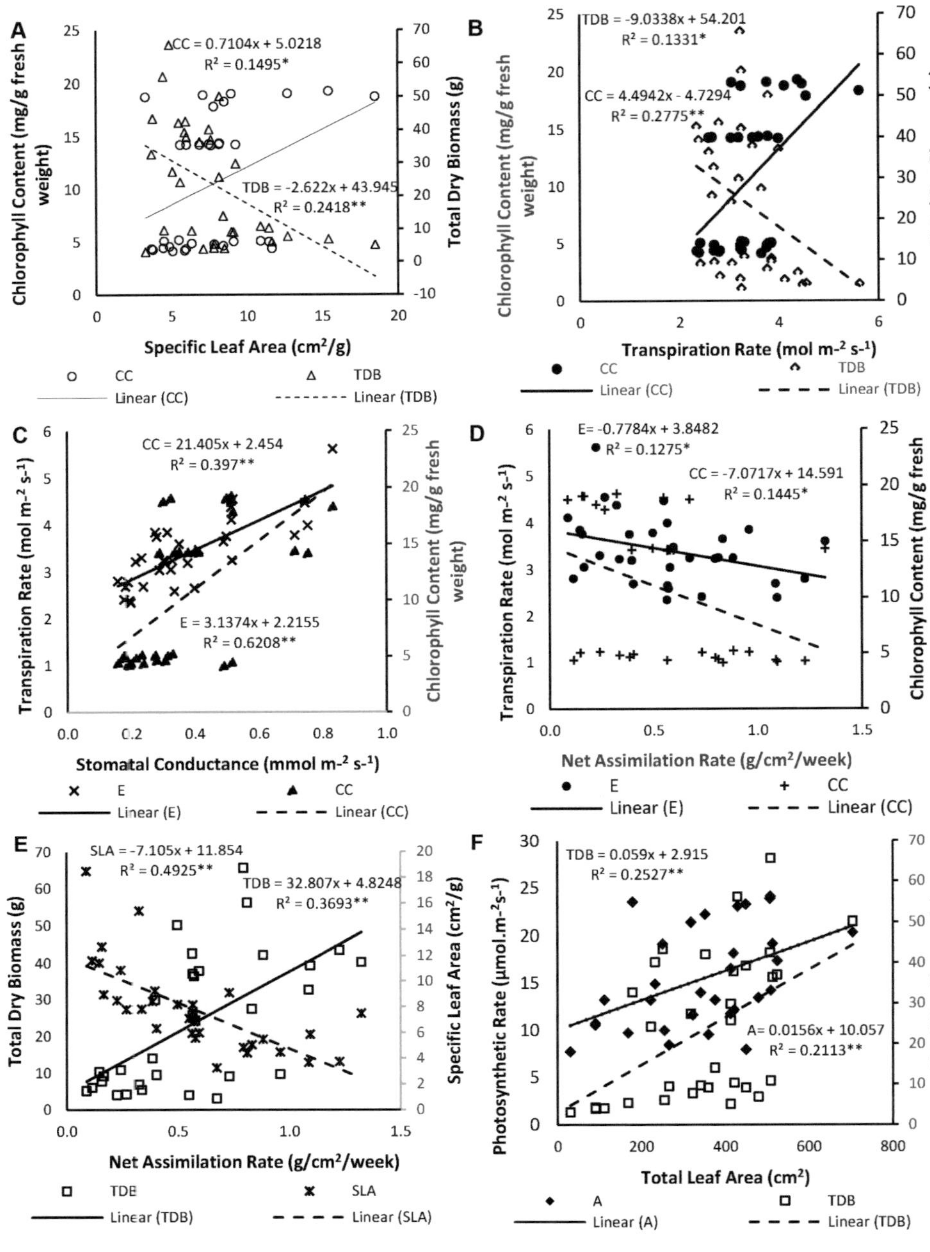

Figure 6.
Correlation coefficient between CC and TDB with (A) SLA; (B) E; E and CC with (C) gs; (D) NAR; TDB and SLA with (E) NAR; (F) TLA. $^{*} = p \leq 0.05$, $^{**} = p \leq 0.01$, $n = 128$.

leaf area with net assimilation rate. Increase in total dry biomass, transpiration rate, chlorophyll content and specific leaf area was associated with a decrease in specific leaf area, transpiration rate and net assimilation rate with an r value of 24.18, 13.31, 12.75, 14.45, and 49.25%, respectively.

10. Conclusions

Brassinolide application had brought notable changes in growth and physiology among fig varieties. Though increasing BL concentration (50, 100 and 200 ml.L^{-1})

caused some differences in growth and physiological changes of fig, but the differences were not consistent and most of the changes happened only in first or second month. Cultivar IBT showed higher growth and physiological changes than cultivar MD after receiving brassinolide treatment. There was significant effect of interaction between brassinolide and variety on growth and physiological changes of fig except for plant height and total dry biomass.

Author details

Zulias Mardinata[1*], Mardaleni[2] and Tengku Edy Sabli[2]

1 Department of Agronomy, School of Graduate Studies, Islamic University of Riau, Marpoyan, Pekanbaru, Indonesia

2 Department of Agrotechnology, Faculty of Agriculture, Islamic University of Riau, Marpoyan, Pekanbaru, Indonesia

*Address all correspondence to: zulias@agr.uir.ac.id

References

[1] Clouse SD, Sasse JM. BRASSINOSTEROIDS: Essential regulators of plant growth and development. Annual Review of Plant Physiology and Plant Molecular Biology. 1998;**49**(1):427-451. DOI: 10.1146/annurev.arplant.49.1.427

[2] Khripach V, Zhabinskii V, De Groot A. Twenty years of brassinosteroids: Steroidal plant hormones warrant better crops for the XXI century. Annals of Botany. 2000;**86**(3):441-447. DOI: 10.1006/anbo.2000.1227

[3] Grove MD, Spencer GF, Rohwedder WK, Mandava N, Worley JF, Warthen JD, et al. Brassinolide, a plant growth-promoting steroid isolated from *Brassica napus* pollen. Nature. 1979;**281**(5728):216-217. DOI: 10.1038/281216a0

[4] Sandalio LM, Rodríguez-Serrano M, Romero-Puertas MC. Leaf epinasty and auxin: A biochemical and molecular overview. Plant Science. 2016;**253**(1):187-193. DOI: 10.1016/j.plantsci.2016.10.002

[5] Vardhini BV. Effect of brassinolide on certain enzymes of sorghum grown in saline soils of Karaikal. The Journal of Phytological Research. 2012;**4**:30-33

[6] Clouse SD. "Brassinosteroid". The Arabidopsis Book. American Society of Plant Biologists; 2002:1-22. DOI: 10.1199/tab. 0009

[7] Caroll J. Fig Types: Different Types of Fig Trees for the Garden. 2015. Available from: https://www.gardeningknowhow.com/edible/fruits/figs/different-types-of-fig-trees.htm

[8] Kumar V, Radha A, Kumar Chitta S. In vitro plant regeneration of fig (*Ficus carica* L. cv. gular) using apical buds from mature trees. Plant Cell Reports. 1998;**17**(9):717-720. DOI: 10.1007/s002990050471

[9] Ahmad K. Types of Fig Fruit that have been Circulating in Indonesia. [Video File]. Retrieved 9 November 2016. Available from: www. youtube.com/watch?v=dw6JFXq YKX0

[10] Mitchell JW, Mandava N, Worley JF, Plimmer JR, Smith MV. Brassins—A new family of plant hormones from rape pollen. Nature. 1970;**225**(5237):1065-1066. DOI: 10.1038/2251065a0

[11] Brosa C, Capdevila JM, Zamora I. Brassinosteroids: A new way to define the structural requirements. Tetrahedron. 1996;**52**:2435-2448. DOI: 10.1016/0040-4020(95)01065-3

[12] Marumo S, Hattori H, Abe H, Nonoyama Y, Munakata K. The presence of novel plant growth regulators in leaves of *Distylium racemosum* Sieb et Zucc. Agricultural and Biological Chemistry. 1968;**32**:528-529. DOI: 10.1271/bbb1961.32.528

[13] Mandava NB. Plant growth-promoting brassinosteroids. Annual Review of Plant Physiology and Plant Molecular Biology. 1988;**39**:23-52. DOI: 10.1146/annurev.pp.39.060188.000323

[14] Iglesias-Arteaga MA. Brassinolide. 2016. Available from: http://www.chm.bris.ac.uk/motm/brassinolide/brassinolideh.htm

[15] Bishop GJ, Yokota T. Plants steroid hormones, brassinosteroids: Current highlights of molecular aspects on their synthesis/metabolism, transport, perception and response. Plant & Cell Physiology. 2001;**42**:114-120. DOI: 10.1093/pcp/pce018

[16] Tang J, Han Z, Chai J. Q&A: What are brassinosteroids and how do they

act in plants. BMC Biology. 2016;**14**:113. DOI: 10.1186/s12915-016-0340-8

[17] Angel R, Chadha YR. The wealth of India. Kew Bulletin. 1974;**29**:496. DOI: 10.2307/4108144

[18] Litz RE, Pliego-Alfaro F, Hormaza JI, editor. Biotechnology of Fruit and Nut Crops. CABI; 2005. pp. 1-65. DOI: 10.1079/9780851996622.0000

[19] Fem K. Edible Perennials: 50 Top Perennials from Plants for a Future. East Meon, United Kingdom: Permanent Publications; 2015. pp. 10-84

[20] Flaishman M, Rodov V, Stover E. Fig (*Ficus carica*): Botany, horticulture and breeding. Horticultural Reviews. 2008;**34**:113-196

[21] Lim TK. Edible Medicinal And Non-Medicinal Plants. Vol. 3. New York: Springer; 2016. pp. 1-659. DOI: 10.1007/978-94-017-7276-1

[22] Condit IJ. Fig varieties: A monograph. Journal of Chemical Information and Modeling. 2013;**53**:1689-1699. DOI: 10.1017/CBO9781107415324.004

[23] Ritchie G, Landis T. The container tree nursery manual. Journal of Colloid and Interface Science. 2010;**404**:161-168. DOI: 10.1016/j.jcis.2013.05.010

[24] Altorenna A. How to Grow Fig Trees in Containers. 2016. Available from: https://dengarden.com/gardening/growing-fig-trees [Retrieved 19 October 2019]

[25] Mayumi H, Shibaoka K. A possible double role for Brassinolide in the reorientation of cortical microtubules in the epidermal cells of Azuki Bean Epicotyls. Plant & Cell Physiology. 1995. DOI: 10.1093/oxfordjournals.pcp.a078734

[26] Prusakova LD, Chizhova SI, Tretyakov NN, Ageeva LF, Golantseva EN, Yakovlev AF. Ecost and epibrassinolide antistress functions on spring wheat under the conditions of the Central Non-Chernozem zone. Agrarian Russia. 1999;**3**:39-41

[27] Wang TW, Cosgrove DJ, Arteca RN. Brassinosteroid stimulation of hypocotyl elongation and wall relaxation in pakchoi (*Brassica chinensis* cv Lei-Choi). Plant Physiology. 1993;**101**(3):965-968. DOI: 10.1104/pp.101.3.965

[28] Hu YJ, Shi LX, Sun W, Guo JX. Effects of abscisic acid and brassinolide on photosynthetic characteristics of Leymus chinensis from Songnen Plain grassland in Northeast China. Botanical Studies. 2013;**54**(1):42. DOI: 10.1186/1999-3110-54-42

[29] Bera A, Pramanik K, Mandal B. Response of biofertilizers and homobrassinolide on growth, yield and oil content of sunflower (*Helianthus annuus* L.). African Journal of Agricultural Research. 2014;**9**:3494-3503

[30] Anjum SA, Wang LC, Farooq M, Hussain M, Xue LL, Zou CM. Brassinolide application improves the drought tolerance in maize through modulation of enzymatic antioxidants and leaf gas exchange. Journal of Agronomy and Crop Science. 2011;**197**:177-185. DOI: 10.1111/j.1439-037X.2010.00459.x

[31] Zaharah S, Razi M, Zainuddin M, Ghani YA. Growth performance and yield of mango in response to restricted rooting volume. Proceeding of Agriculture Congress. 2006. p. 365-371

[32] Henson IE. Carbon assimilation, respiration and productivity of young oil palm (*Elaeis guineensis*). Elaeis. 1992;**4**:51-59

[33] Haniff MH. Gas exchange of excised oil palm (*Elaeis guineensis*) fronds. Asian Journal of Plant Sciences.

2006;**5**(1):9-13. DOI: 10.3923/ajps.2006.9.13

[34] Lambers H, Poorter H. Inherent variation in growth rate between higher plants: A search for physiological causes and ecological consequences. Advances in Ecological Research. 1992;**2504**(08):60148. DOI: 10.1016/S0065-2504(08)60148-8

[35] Hayat S, Alyemeni MN, Hasan SA. Foliar spray of brassinosteroid enhances yield and quality of *Solanum lycopersicum* under cadmium stress. Saudi Journal of Biological Sciences. 2012;**19**(3):325-335. DOI: 10.1016/j.sjbs.2012.03.005

[36] Gardner FP, Pearce RB, Mitchell RL. Physiology of Crop Plants. Jodhpur, India: Scientific Publishers; 2017

[37] Rajput A. Physiological parameters leaf area index, crop growth rate, relative growth rate and net assimilation rate of different varieties of rice grown under different planting geometries and depths in SRI. Indian Journal of Pure & Applied Biosciences. 2017;**5**(1):362-367. DOI: 10.18782/2320-7051.2472

[38] Lin X. Physiological effect and yield increase action after spraying BR in rice early blooming stage. Journal of Anhui Agricultural Sciences. 2007;**35**:3317

[39] Chen L, Li Y, Zheng F, Ren Z, Guan R, Liu Z, et al. Effect of brassionsteroids on soybean resistance to *Phytophthora sojae*. Soybean Science. 2008;**26**:713

[40] Shu H, Guo S, Shen X, Ni W. Cotton physiology affected by brassinosteroid under NaCl stress. Jiangsu Journal of Agricultural Sciences. 2011;**27**:1198-1202

[41] Bao F, Shen J, Brady SR, Muday GK, Asami T, Yang Z. Brassinosteroids interact with auxin to promote lateral root development in arabidopsis. Plant Physiology. 2004;**134**(4):1624-1631. DOI: 10.1104/pp.103.036897

[42] Sairam RK. Effects of homobrassinolide application on plant metabolism and grain yield under irrigated and moisture-stress conditions of two wheat varieties. Plant Growth Regulation. 1994;**14**(2):173-181. DOI: 10.1007/BF00025220

[43] Iwahori S, Tominaga S, Higuchi S. Retardation of abscission of citrus leaf and fruitlet explants by brassinolide. Plant Growth Regulation. 1990;**9**(2):119-125. DOI: 10.1007/BF00027439

[44] Ibrahim MH, Jaafar HZE, Rahmat A, Rahman ZA. The relationship between phenolics and flavonoids production with total non structural carbohydrate and photosynthetic rate in *Labisia pumila* Benth. under high CO_2 and nitrogen fertilization. Molecules. 2011;**16**(1):162-174. DOI: 10.3390/molecules16010162

Chapter 3

The Creation of Resistant Berries' Agrobiocenosis

Zoya Evgenievna Ozherelieva, Pavel Sergeevich Prudnikov, Diana Aleksandrovna Krivushina, Marina Ivanovna Zubkova and Anna Androsova

Abstract

In *conditions* of increasing differences in the hydrothermal regime of the environment, all known adaptation mechanisms should be used fully, as at the level of individual varieties and at agrobiocenosis in general. The modern gardener needs high-productive, adapted to the growing conditions varieties of the *Fragária × ananássa* Durh. The resistant varieties of the *Fragária × ananássa* Durh. to negative environmental conditions is a basic characteristic, which shows an economic value and the effectiveness of their cultivation in an actual zone. The main physiological and biochemical indicators of resistance during autumn hardening and after temperature stress in winter are changes in the antioxidant system, interruptions of the protein-carbohydrate complex, accumulation products of membran's lipoperoxidation, and changes in the fractional composition of water in leaves. The *Fragária × ananássa* Durh. production process is characterized by the following physiological parameters: pigment analysis, photochemical activity of isolated chloroplasts, respiration, and net photosynthesis productivity. Studies of physiological and biochemical resources of resistance to abiotic stress factors and productivity of *Fragária × ananássa* Durh. are shown. As a result of the carried studies, perspective variety of the *Fragária × ananássa* Durh. ("Tzaritza") was identified for the creation of resistance berries' agrobiocenosis.

Keywords: agrobiocenosis, Fragária × ananássa Durh., abiotic stress factors, frost-resistant, resistance to spring frosts, production process

1. Introduction

Truly, strawberries are on the first place among berry crops in the world, due to their excellent taste, attractive appearance, and early fruit maturation [1–3].

The constant introduction of this culture from different climatic zones conductive to the expansion of the assortment, the involvement of new genotypes in the selection process [4]. But often the most productive, large-fruited industrial varieties have low winter hardiness. The realization of garden crops' stability and the intensity of their production processes are significantly determined by their adaptability and their ability to use fully the bioclimatic potential of the placement zone [5, 6]. Resistance to low temperatures and average daily temperature changes are the most important characteristics of the strawberry's variety in the central region [3, 7, 8]. The strawberries die in snowless winters when the temperature decrease from −15 to −18°C but can tolerate temperatures from −25 to −35°C when the level of snow cover at least 20–30 cm. The most dangerous periods in overwintering—snowless late autumn and early winter—November, December, when the snow has not yet fallen, and the air temperature decrease to −10, −15°C, can be possible freezing or death of the root system of strawberries [9]. At this time, the leaves and flower buds can freeze and in more intensive frosts—branch crowns and whole bushes. Especially the low winter-resistant varieties have suffer distress and plants that are prepared for winter badly. The second critical period for strawberries is the end of winter—the beginning of spring. The snow on the plantations is beginning to settle down and melt, and the bushes are opening [10, 11]. Alternation of thaws and frosts is also dangerous for plants [12], when the snow melts near the ground and the snow crust remains on top. At this time, there is a getting wet of bushes [10]. Visible damage after low temperatures is the death of a whole plant or damage the branch crowns and rhizome.

The frost-resistant state of plants is achieved under the condition of stopping growth and passing through the hardening phases [13]. The great significance of hardening for successful overwintering of strawberries is revealed [14]. The hardening is a difficult complex of physiological and biochemical changes which is associated with some cell dehydration and with the accumulation of protective compounds (sugars, low-molecular water-soluble proteins, amino acids, etc.), which, by increasing the concentration of cellular fluid and binding free water in the plant, prevent the formation of intracellular ice. At critical temperatures, the water outflow from the cells becomes worse significantly, and a lot of supercooled water appears, which then freezes inside the protoplast and can lead to cell death [15]. Thus, plants are characterized by an increase in the amount of bound water by the beginning of the winter period [16]. The ratio of free water to bound water is one of the essential signs of plant adaptation to a temperature decrease in the autumn-winter period. It is noted that the ratio of free water to bound water is lower in winter-hardy varieties than in non-winter-hardy ones. The state of the water regime in autumn and the effective accumulation of protective substances by winter are the important factors, which determine the successful overwintering of plants [16–18].

The research was made at the section of primary variety research of VNIISPK. The strawberry plants were planted in 2016, in the second half of summer (end of July-start of August) according to the scheme 90 × 20 cm in threefold replication, on 30 plants in every replication, randomized. We studied varieties of strawberries of different ecological and geographical origins ("Kokinskaya rannyaya," "Solovushka," "Rosinka," "Tsaritsa," "Urozhainaya TzGL" [Russia]; "Sara" [Sweden]; "Alba," "Marmolada" [Italy]; "Korona," "Sonata" [Holland]) to identify the features of functional conjugation of physiological and biochemical processes of resistance to the action of low-temperature environmental factors and productivity. **Table 1** shows the minimum and maximum air temperature in the autumn and early winter during the years of project realization.

T (°C)	2017				2018			
	Sept	Oct	Nov	Dec	Sept	Oct	Nov	Dec
Max t°C	28.0	15.5	9.0	9.0	29.8	21.5	10.6	1.5
Min t°C	−1.5	−4.8	−9.2	−5.5	−1.0	−2.8	−18.5	−17.0

Table 1.
Air temperature in the period September-December 2017–2018.

2. The study of the physiology-biochemical parameter resistance of varieties of strawberry from different ecological and geographical origins in autumn period

As known, the winter-hardy varieties of fruit crops have a higher bound/free water ratio, than non-winter-hardy [6]. In September, when the fractional composition of water was determining, the strawberry leaves showed a low ratio of bound water to free water. In October, this ratio increased by 1.8–7.7 times. In November, as far as the air temperature decreased (**Table 1**), the bound/free water ratio in leaves was higher (1.6–18.0 times) than in the previous autumn months. The largest measure of the water fraction ratio to the beginning of winter was noted in the varieties—"Solovushka," "Tsaritsa," "Sara," and "Korona" (**Figure 1**). Correlation analysis showed a high level of dependence between the minimum air temperature in autumn, the water content of strawberry leaves tissues (r = 0.81), and the bound/free water ratio (r = −0.97).

According to some of the researchers, proline has osmoprotective properties under stressful conditions [19–21]. In September-November period of time, as far as the air temperature decreased (**Table 1**), an increase in the amount of the amino acid proline was noted in the leaves. During cold adaptation, an increase in the amount of proline in strawberry plants was registered by other authors [22]. In our studies, the most intense peak of proline accumulation in all varieties was noted in October (3.27–11.21 times compared to September), while in November (1.19...1.91 times compared to October) (**Table 2**). At the same time, the varieties "Solovushka," "Kokinskaya rannyaya," and "Tsaritsa" were characterized by the highest level of the amino acid proline accumulation. The amount of proline increased by 10.38–16.20 times at these varieties in November compared to September.

Low-molecular carbohydrates play an important role as osmoprotectors alongside with proline [23]. Sugars increase the water-holding ability of protoplasmic colloids, protecting them from ice formation and excessive cell dehydration. In our studies, during 2 years, significant accumulation of sugars was noted in strawberry leaves in October compared to September, on the background of a decrease in air temperature (**Table 3**). So in 2017, the level of carbohydrates in leaves tissues increased by 1.92–4.93 times and in 2018 by 1.20–3.34 times. The maximum accumulation of sugars in 2017 was marked at varieties of strawberries "Sara," "Korona," "Tsaritsa," and "Urozhainaya TzGL" (the amount of carbohydrates in October increased by 2.37–4.93 times compared to September). At "Solovushka"

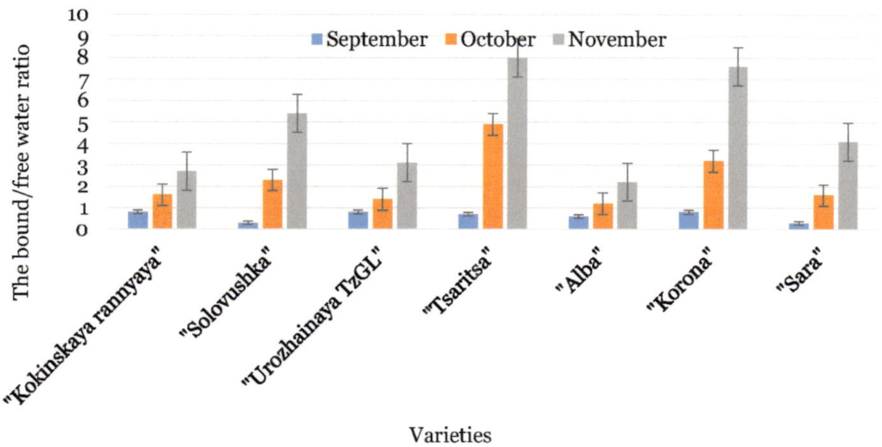

Figure 1.
The bound/free water ratio in the leaf's tissues of strawberry varieties in the autumn period.

Varieties	Proline, mg/kg		
	September	October	November
"Kokinskaya rannyaya"	2.34 ± 0.12	14.95 ± 0.79	24.28 ± 1.33
"Solovushka"	1.19 ± 0.06	13.34 ± 0.67	19.38 ± 0.90
"Urozhainaya TzGL"	1.70 ± 0.11	12.48 ± 0.68	14.93 ± 0.79
"Tsaritsa"	2.65 ± 0.11	11.32 ± 0.40	28.73 ± 1.15
"Alba"	2.03 ± 0.11	11.80 ± 0.59	14.90 ± 0.74
"Korona"	3.95 ± 0.15	14.34 ± 0.49	27.65 ± 1.11
"Sara"	6.07 ± 0.30	19.87 ± 098	27.55 ± 1.38

Table 2.
The content of free proline in the leaves of strawberry plants in autumn.

and "Kokinskaya rannyaya" varieties, the amount of sugars increased at a lower level in compared to September, by 1.95 and 2.10 times. However, in 2018, in October compared to September, the maximum level of sugar accumulation was remained only at the "Tsaritsa" variety—the amount of carbohydrates increased by 3.34 times. In addition, a high movement of sugar biosynthesis was marked at the "Solovushka" variety—by 2.20 times as compared to September. The low sugar accumulation was at the "Urozhainaya TzGL," "Tsaritsa," and "Alba" varieties by 1.20–1.57 times, in October 2018. In November 2017, as the temperature decreased, some varieties showed a decrease in the amount of sugars compare to the October level, which can be ascribed to their active use as an energy substrate for respiration processes, protein synthesis, amino acids, etc. At the same time, the sugar content decreased to a largest level at the varieties "Urozhainaya TzGL," "Sara," and "Alba" (by 1.31–1.57 times compared to October). However, sugar content continued to increase at the "Tsaritsa" and "Korona" varieties. The changes in the level of sugars were not significant at "Solovushka" and "Kokinskaya rannyaya" varieties (**Table 3**).

The determining of the total protein in leaves of strawberry showed the significant increase in its amount in October to November (**Figure 2**), which can explain the decrease in the amount of low-molecular carbohydrates observed in some varieties in November. At the background of high accumulation of proline, the "Solovushka," "Tsaritsa," "Korona," and "Sara" varieties were characterized not only by more active protein biosynthesis (the amount of peptide compounds increased by 2.8–3.5 times in November compared to October) but also had the highest protein content in November.

When resistance is forming, the functioning of the antioxidant protection system is significant, which prevents the development of oxidative stress and, in particular, peroxidation of membrane lipids (POL) at the background of adverse environmental factors. The intensity of damage to cell membranes was estimated by the accumulation of the final product of lipid peroxidation, malondialdehyde (MDA); a transition of lipid peroxidation, hydroperoxides; the content of hydrogen peroxide (as a representative of reactive oxygen species); and the activity of antioxidant enzymes: superoxide dismutase (SOD), catalase, and peroxidase.

The results of the study showed that as far as temperature of environment was decreasing, the intensity of MDA accumulation increased in all varieties. However, the intensity of damage to membrane lipids was significantly lower at "Solovushka," "Sara," "Korona," and "Tsaritsa" varieties, than in other genotypes. As can be seen from **Table 4**, the content of MDA increased by 17.5–25.7% at "Solovushka," "Sara,"

Varieties	Sucrose, mg/g					
	2017		2018			
	September		October		November	
"Kokinskaya rannyaya"	1.32 ± 0.07	3.92 ± 0.24	2.57 ± 0.11	5.94 ± 0.30	2.31 ± 0.11	8.54 ± 0.56
"Solovushka"	1.30 ± 0.06	4.97 ± 0.27	2.73 ± 0.15	10.91 ± 0.65	2.65 ± 0.13	9.95 ± 0.56
"Urozhainaya TzGL"	1.54 ± 0.08	3.85 ± 0.23	3.65 ± 0.15	4.63 ± 0.21	2.23 ± 0.10	7.78 ± 0.45
"Tsaritsa"	1.43 ± 0.05	4.68 ± 0.27	7.05 ± 0.28	15.62 ± 0.70	8.34 ± 0.29	16.56 ± 1.08
"Alba"	1.78 ± 0.09	4.24 ± 0.21	3.41 ± 0.17	6.60 ± 0.38	2.04 ± 0.11	7.35 ± 0.44
"Korona"	1.15 ± 0.05	3.84 ± 0.24	4.16 ± 0.15	7.05 ± 0.28	6.46 ± 0.26	8.90 ± 0.48
"Sara"	0.98 ± 0.05	3.37 ± 0.21	3.50 ± 0.14	6.26 ± 0.36	2.68 ± 0.12	8.00 ± 5.14

Table 3.
The content of sucrose in the leaves of strawberry plants in autumn.

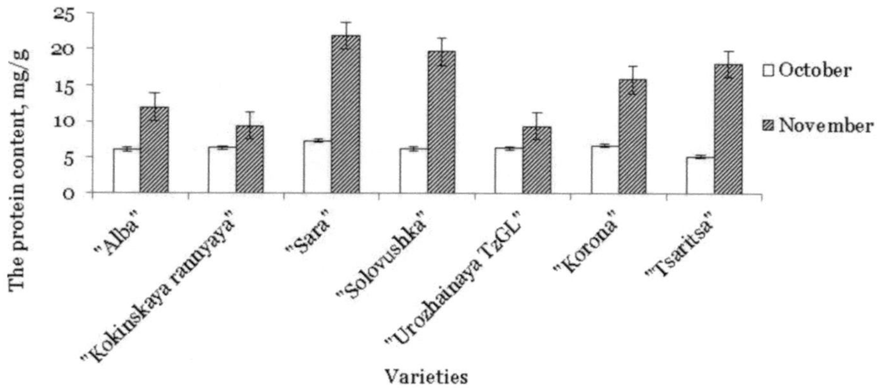

Figure 2.
The content of total protein in the leaves of strawberry plants in autumn.

"Korona," and "Tsaritsa", in November 2017 compared to October, while in other varieties it increased by 31.7–56.3%. At the same time, absolute MDA readings at "Solovushka," "Sara," "Korona," and "Tsaritsa" were lower in November than other cultivars, which indicate about a more significant damage of the structural-functional integrity of cell membranes in other varieties (**Table 4**). Different levels of MDA accumulation in the studied varieties seem to be associated with different degrees of formation of reactive oxygen species in cells and with higher activity of the antioxidant defense system that neutralizes reactive oxygen species (ROS). In this regard, it became necessary to define hydrogen peroxide as one of the representatives of ROS.

Thus, a correlation between MDA and H_2O_2 was shown, when determining hydrogen peroxide. At the same time, the dependence between these indicators changed over the years. So, the correlation coefficient between the amount of MDA and the content of hydrogen peroxide in tissues was, in October 2017, r = 0.67 and, in November, r = 0.79. In 2018, the correlation between the level of H_2O_2 and the intensity of POL in October was 0.55, but in November it increased to 0.98, which is explained by a significant decrease in the last month of autumn air temperature (**Table 1**) and the development of oxidative stress. At the varieties "Solovushka," "Korona," "Tsaritsa," and "Sara", not only the level of accumulation of hydrogen

Varieties	MDA, microMol/g			
	2017		2018	
	October	November	October	November
"Kokinskaya rannyaya"	4.1 ± 0.14	5.4 ± 0.23	10.1 ± 0.56	15.8 ± 0.84
"Solovushka"	3.8 ± 0.28	4.5 ± 0.25	9.8 ± 0.54	10.6 ± 0.48
"Urozhainaya TzGL"	3.9 ± 0.12	5.8 ± 0.17	8.1 ± 0.36	11.9 ± 0.62
"Tsaritsa"	3.7 ± 0.13	4.5 ± 0.18	7.5 ± 0.34	8.8 ± 0.37
"Alba"	4.8 ± 0.17	7.5 ± 0.15	11.4 ± 0.51	21.0 ± 1.30
"Korona"	3.5 ± 0.12	4.4 ± 0.20	8.5 ± 0.37	10.3 ± 0.46
"Sara"	4.0 ± 0.12	4.7 ± 0.20	9.5 ± 0.43	11.2 ± 0.61

Table 4.
The content of MDA in the leaves of strawberry plants in autumn.

peroxide was significantly lower, but also the absolute values of H_2O_2 were reduced in comparison with the other genotypes in October and November (**Table 5**).

Analysis of the antioxidant enzymes activity showed their significant intensity at varieties with low levels of MDA, hydroperoxides, and hydrogen peroxide, with some exceptions for superoxide dismutase (SOD). Thus, in the "Solovushka," "Tsaritsa," "Korona," and "Sara" varieties, the activity of SOD, an enzyme that recycles superoxide with the formation of hydrogen peroxide, did not significantly change in November compared to October during 2 years of research, while in the other genotypes, it increased by 14.6–31.4% (**Table 6**). On the one hand, this explains the different levels of hydrogen peroxide and hydroperoxides in the studied varieties.

In addition to the study of the activity of another antioxidant enzyme, catalase showed its significant intensification in "Solovushka," "Tsaritsa," "Korona," and "Sara" varieties. The activity of hydrogen peroxide scission by catalase at these varieties increased by 37.7–50.5% in November 2017 compared to October, while in the others by 15.7–25.5%. The correlation coefficient between the activity of the enzyme and the level of hydrogen peroxide, which is involved in lipoperoxidation of cell membranes, was r = −0.20 in October and in November r = −0.71. In 2018, the dependence between the amount of H_2O_2 and the enzymes' activity was stronger and was r = −0.84 in October and r = −0.82 in November. At the same time, the varieties "Solovushka," "Tsaritsa," "Korona," and "Sara" were characterized by a large increase in catalase activity in November 2018 (**Table 7**).

The correlation analysis showed the high level of dependence between the physiology-biochemical parameters of strawberries and the minimum air temperature in autumn (**Table 8**). In addition, the significant dependence was between bound/free water ratio, the content of proline (r = 0.98–0.99), and the sucrose (r = 0.72–0.97).

So, the multifaceted study of the physiology-biochemical parameter resistance of strawberry varieties from different ecological and geographical origins was made in autumn period. As a result of this experiment, it was found that the increase of bound water and decrease of free water in leaves were characterized in autumn period for strawberry plants on the background decrease of water content level. Changes in the composition of water fractions depended largely on the accumulation of sucrose and free proline in the leaves of strawberry during the autumn period. A high dependence between the physiology-biochemical parameters and the minimum air temperature was established during the autumn adaptation of strawberries. At the same time, "Solovushka," "Tsaritsa," "Sara," "Korona," varieties

Varieties	Hydrogen peroxide, microMol/g			
	2017		2018	
	October	November	October	November
"Kokinskaya rannyaya"	3.1 ± 0.15	15.7 ± 0.32	3.3 ± 0.18	14.8 ± 0.99
"Solovushka"	1.5 ± 0.05	3.5 ± 0.12	1.8 ± 0.10	4.2 ± 0.24
"Urozhainaya TzGL"	2.2 ± 0.17	7.4 ± 0.30	2.9 ± 0.19	9.5 ± 0.58
"Tsaritsa"	1.6 ± 0.06	2.7 ± 0.09	2.0 ± 0.13	3.6 ± 0.20
"Alba"	2.9 ± 0.15	16.9 ± 0.21	3.5 ± 0.23	21.2 ± 1.42
"Korona"	1.7 ± 0.06	2.9 ± 0.11	2.2 ± 0.14	5.3 ± 0.32
"Sara"	1.9 ± 0.10	3.9 ± 0.12	2.1 ± 0.13	4.6 ± 0.25

Table 5.
The content of hydrogen peroxide (H_2O_2) in the leaves of strawberry plants in autumn.

Varieties	SOD, c.u.			
	2017		2018	
	October	November	October	November
"Kokinskaya rannyaya"	71.0 ± 0.86	83.5 ± 3.21	87.1 ± 5.67	111.2 ± 7.23
"Solovushka"	55.5 ± 1.20	54.2 ± 1.25	65.4 ± 3.60	66.2 ± 3.97
"Urozhainaya TzGL"	69.2 ± 1.91	79.3 ± 1.29	86.7 ± 3.90	106.6 ± 5.12
"Tsaritsa"	53.4 ± 1.87	52.5 ± 1.78	65.5 ± 3.93	67.3 ± 4.03
"Alba"	53.6 ± 1.64	65.5 ± 3.04	74.3 ± 4.09	97.6 ± 5.86
"Sara"	66.9 ± 2.39	63.1 ± 2.06	70.5 ± 3.88	68.4 ± 4.10
"Korona"	52.7 ± 1.84	51.5 ± 1.55	72.3 ± 4.34	77.2 ± 4.86

Table 6.
The SOD activity in the leaves of strawberry plants in autumn.

Varieties	Catalase, ml O_2/min.			
	2017		2018	
	October	November	October	November
"Kokinskaya rannyaya"	9.8 ± 0.36	12.3 ± 0.42	7.5 ± 0.30	8.6 ± 0.34
"Solovushka"	10.5 ± 0.50	15.2 ± 0.36	9.5 ± 0.52	14.5 ± 0.87
"Urozhainaya TzGL"	10.8 ± 0.46	12.5 ± 0.40	9.1 ± 0.45	10.6 ± 0.45
"Tsaritsa"	10.7 ± 0.40	16.1 ± 0.44	12.2 ± 0.72	19.3 ± 0.87
"Alba"	11.7 ± 0.35	14.1 ± 0.47	7.2 ± 0.39	9.1 ± 0.46
"Korona"	11.4 ± 0.44	15.8 ± 0.50	10.3 ± 0.49	14.9 ± 0.77
"Sara"	12.2 ± 0.35	16.8 ± 0.50	10.2 ± 0.53	13.3 ± 0.69

Table 7.
The catalase activity in the leaves of strawberry plants in autumn.

had the highest bound/free water ratio, less damage of the structural and functional integrity of cell membranes, a low level of accumulation of hydrogen peroxide and hydroperoxides, and an increase in the activity of the antioxidant enzyme catalase,

The physiology-biochemical parameters	The correlation coefficient, r
Water content of leaves	0.81
The bound/free water ratio	−0.97
Proline	−0.98
Sucrose	−0.70
MDA	−0.92
H_2O_2	−0.82
Hydrogen peroxide	0.90
The catalase activity	−0.94
The SOD activity	−0.98

Table 8.
The correlation coefficient between the physiology-biochemical parameters of strawberry leaves and the minimum air temperature in the autumn period (min t).

which indicates a greater adaptive ability to low-temperature stresses. The analysis of another antioxidant enzyme—peroxidase—and the increase of cyanidin level were less informative for strawberry plants in forming resistance to temperature drop in the autumn period.

3. Winter hardiness of strawberries in the field

Over the years of research, we analyze the extent of winter damage to varieties of strawberries of different ecological and geographical origins, in the field. The characteristic of the variety for winter hardiness is determined on the basis of spring accounting of the degree of freezing (in points): degree of freezing by plant regrowth (April); in severe winters – by rhizome freezing [24].

The winter 2017–2018 was mild and snowy. Sharp fluctuations in temperature without snow cover were not observed, which had a favorable effect on overwintering. In January, the minimum air temperature decreased to −15.0°C and on the snow surface −12.5°C. The height of the snow cover at the end of January was 13 cm. In February, we observed a decrease in the minimum air temperature to −26.0 and – 18.5°C on the snow surface. The height of the snow cover reached −15 cm.

The winter 2018–2019 was frosty but snowy. Sharp fluctuations in temperature without snow cover are not observed too, which had a favorable impact on the overwintering of strawberry plants. In January, the minimum air temperature decreased to −24.5°C and on the snow surface −21.0°C. At the end of January, the height of the snow cover was 41 cm. In February, we observed a decrease in the minimum air temperature to −11.5 and – 11.5°C on the snow surface. The height of the snow cover reached 35 cm. So, most of the strawberry varieties survived without damage in the winter conditions of 2017–2018 and 2018–2019. The degree of freezing of most samples was 0.0 points, and some varieties had minor damage ("Sonata" and "Marmolada" up to 1.0 points). Also, the estimation of common generation of plants is demonstrative, and it was conducted in late May and early June in points. This characteristic depends largely from the hardiness and shows how the plants overwintered and in what state they enter the fruiting phase. The common condition at the varieties "Kokinskaya rannyaya," "Rosinka," "Solovushka," "Urozhainaya TzGL," "Tsaritsa," and "Sara" at the beginning and at the end of the growing season was excellent (5.0 points). The common condition of plants at the end of vegetation was excellent—80% at the most varieties. So, the weather conditions of the winter period 2017–2019 were favorable for strawberry plants overwintering.

4. Frost resistance at the beginning of the winter without snow cover and reaction on thaws of strawberry plants in winter period

Artificial freezing makes it possible to screen fruit and berry plants to determine the winter hardiness biopotential [25, 26]. In low-snow and snowless winters, the branch crowns of strawberry plants are slightly damaged by negative temperatures, but the strong freezing of the rhizomes is noted on this, as a result of this, the growth and development of plants are delayed. To determine the resistance of varieties of strawberries to early winter frost, we used freezing modes at the end of November (−15°C) and in early December (−20°C).

After exposure of a temperature −15°C, the plants of the varieties "Rosinka" and "Tsaritsa" were without damages. Insignificant damages (no more than 1.0 points) were seen in strawberry varieties "Solovushka," "Korona," and "Sara" on the top of the rhizome, as a very weak browning. The plants quickly recovered and developed well, when they were growing. Average freezing of rhizomes was detected in varieties "Alba" and "Marmolada." The rhizome tissue was light brown. The branch crowns of these varieties grew more slowly, than those of others, but later they developed normally (**Table 9**).

After an exposure of a temperature of −20°C, in early December, insignificant damages were noted at "Solovushka," "Rosinka," and "Tsaritsa" varieties (the degree of damage is not more than 1.0 points, **Figure 3**). At the same time, varieties with reversible damage to the branch crown tissues ("Korona", "Sara") were identified. Average freezing of rhizomes was detected at the varieties "Kokinskaya rannyaya," "Urozhainaya TzGL," "Alba," and "Sonata." The rhizome tissues were light brown. In the plants of these varieties, uneven growth of branch crowns was noted, but later they recovered and developed normally. The rhizomes of "Marmolada" plants are

Varieties	Point of damage to strawberries in the early winter period	
	−15°C	−20°C
"Urozhainaya TzGL" (st)	1.1 ± 0.10	2.5 ± 0.33
"Kokinskaya rannyaya"	0.7 ± 0.27*	2.5 ± 0.44
"Solovushka"	0.2 ± 0.15**	0.8 ± 0.36***
"Rosinka"	0.0 ± 0.00**	0.3 ± 0.12***
"Tsaritsa"	0.0 ± 0.00**	0.8 ± 0.20***
"Alba"	2.6 ± 0.18***	3.0 ± 0.41
"Marmolada"	2.3 ± 0.20*	3.5 ± 0.31**
"Korona"	0.5 ± 0.18	2.0 ± 0.14
"Sara"	0.8 ± 0.18	2.0 ± 0.33
"Sonata"	1.0 ± 0.00	2.3 ± 0.12

*Significantly at a significance level of $p < 0.05$.
**Significantly at a significance level of $p < 0.01$.
***Significantly at a significance level of $p < 0.001$.

Table 9.
The degree of damage to plant varieties strawberry in the beginning of winter (2017–2019).

a b

Figure 3.
The damage of branch crown tissues by a temperature of −20°C in early December, at varieties "Rosinka" (a) up to 0.5 point and "Solovushka" (b) up to 1.0 point.

very frozen. The tissues of the rhizome variety were dark brown. During regrowth, there was an inhibition in the growth of plants that died in the end (**Table 9**).

The high degree of dependence came between the bound/free water ratio and the degree of damage to strawberry plants at the beginning of winter (r = −0.76). At the same time, a moderate association was determined between the degree to plant strawberry damage, the content of free proline (r = −0.34) and sucrose (r = −0.39) in leaves tissues.

To identify the influence of positive temperature on the frost resistance of strawberry varieties, we simulated a 3-day thaw of +5°C followed by freezing at temperatures of −10°C and − 15°C in December, January, and February.

When the temperature decreased to −10°C, after a 3-day thaw +5°C in December, we noted minor damages to the apical buds of brunch crowns no more, than 1.0 point at some varieties ("Solovushka," "Urozhainaya TzGL," "Korona," "Marmolada," "Sara," "Sonata"). The "Alba" variety showed average freezing of rhizome, and its plants grew more slowly. A high regeneration of the apical buds of branch crowns was noted at most varieties, and the plants developed well. In January, after this freezing model, the damage of degree to the apical buds was no more than 1.6 points at the "Solovushka," "Tsaritsa," "Korona," "Sara," and "Sonata" varieties. The plants of their varieties recovered well after freezing. At the same time, the frost resistance was decreased at "Kokinskaya rannyaya," "Rosinka," and "Urozhainaya TzGL" varieties, which have noted average freezing of the rhizome. The "Marmolada" variety had significant freezing, and the rhizomes were brown. Plants of this variety were weakened and developed poorly. In February, when the temperature decreased to −10°C after a 3-day thaw of +5°C, the varieties "Solovushka," "Tsaritsa," "Korona," and "Sara" had reversible damages to the rhizome (the degree of damage was not more than 2.0 points). The average level of frost resistance during this period was kept by the varieties "Solovushka," "Alba," and "Sonata". Strong freezing was detected at the varieties "Kokinskaya rannyaya,"

"Urozhainaya TzGL," and "Marmolada," in which the branch crown did not grow and the plants died (**Table 10**).

The decrease of temperature to −15°C, after a 3-day thaw +5°C in December, increased freezing at strawberry varieties. The damage of rhizomes and annual branch crowns was not more than 2.0 points at the "Rosinka," "Solovushka," "Tsaritsa," "Korona," and "Sara" varieties. In December, the "Alba" and "Marmolada" varieties froze strongly. In January and February, under the same regime, plants of the "Solovushka," "Tsaritsa," and "Korona" varieties had reversible damage (no more than 2.0 points). Plants have recovered well after freezing. Average frost resistance was noted in the "Sara" variety. Plants of this variety grew slowly. In January and February, the varieties "Kokinskaya rannyaya," "Rosinka," "Urozhainaya TzGL," "Marmolada," and "Sonata" had strong freezing of plants. The plants of this varieties were dead after the further regrowth. In January and February, plants of the "Alba" variety froze strongly and died, after the thaw +5°C and a sharp decrease of temperature to −15°C. As a result, it was shown that the reaction of strawberry varieties to negative temperatures after the thaw increased by the end of the winter period. The most stable frost resistance during the winter thaw was shown by the varieties of "Solovushka," "Tsaritsa," and "Korona" (**Table 11**).

So, the high dependence was noted between the degree of strawberry plants freezing and the bound/free water ratio at the beginning of winter. In the early winter period, varieties of strawberries "Solovushka," "Rosinka," "Tsaritsa," "Korona," and "Sara" were characterized by frost resistance. Also, it was shown that the reaction of strawberry varieties to negative temperatures after the thaw increased towards the end of the winter period. The ability to consistently keep frost resistance during the winter thaw, showed the "Solovushka", "Tsaritsa", "Korona" varieties. As a result of artificial freezing, frost-resistant varieties of strawberries—"Solovushka," "Tsaritsa," and "Korona"—were identified. These results contributed to the development of methodological recommendation for determining the frost resistance of strawberries under controlled conditions.

Varieties	**Point of damage to strawberries during the winter thaw**		
	December	**January**	**February**
	+5, −10°C	**+5, −10°C**	**+5, −10°C**
"Urozhainaya TzGL" (st)	1.3 ± 0.25	2.5 ± 0.96	3.8 ± 0.46
"Kokinskaya rannyaya"	1.8 ± 0.18**	2.5 ± 0.96	3.2 ± 0.38
"Solovushka"	0.6 ± 0.24	1.4 ± 0.55*	1.9 ± 0.23***
"Rosinka"	1.8 ± 0.12**	2.4 ± 1.03	2.5 ± 0.20**
"Tsaritsa"	1.3 ± 0.12	1.6 ± 0.63	1.9 ± 0.13***
"Alba"	2.3 ± 0.27***	2.7 ± 0.62	3.0 ± 0.15
"Marmolada"	0.9 ± 0.13	3.3 ± 0.75	3.5 ± 0.29
"Korona"	1.0 ± 0.00	1.3 ± 0.48*	1.9 ± 0.29***
"Sara"	0.8 ± 0.12	0.8 ± 0.48***	2.0 ± 0.35***
"Sonata"	0.8 ± 0.12	1.6 ± 0.63	2.8 ± 0.32*

**Significantly at a significance level of $p < 0.05$.*
***Significantly at a significance level of $p < 0.01$.*
****Significantly at a significance level of $p < 0.001$.*

Table 10.
The degree of damage to strawberry plants during the thaw +5 and following temperature decreases to −10°C (2017–2019).

Varieties	Point of damage to strawberries during the winter thaw		
	December	January	February
	+5, −15°C	+5, −15°C	+5, −15°C
"Urozhainaya TzGL" (st)	3.0 ± 0.21	4.0 ± 0.58	3.5 ± 0.29
"Kokinskaya rannyaya"	2.7 ± 0.12	3.8 ± 0.43	3.3 ± 0.17
"Solovushka"	1.6 ± 0.15**	1.9 ± 0.08***	2.0 ± 0.58***
"Rosinka"	2.0 ± 0.00*	3.5 ± 0.61	3.0 ± 0.50
"Tsaritsa"	1.7 ± 0.12*	2.0 ± 0.21***	2.0 ± 0.29***
"Alba"	3.1 ± 0.24	4.2 ± 0.27	4.3 ± 0.33*
"Marmolada"	3.4 ± 0.22	3.9 ± 0.08	4.0 ± 0.29
"Korona"	1.3 ± 0.12**	2.2 ± 0.20***	1.9 ± 0.17***
"Sara"	2.0 ± 0.10*	2.4 ± 0.08***	2.5 ± 0.17*
"Sonata"	3.0 ± 0.00	3.8 ± 0.25	3.8 ± 0.18

*Significantly at a significance level of $p < 0.05$.
**Significantly at a significance level of $p < 0.01$.
***Significantly at a significance level of $p < 0.001$.

Table 11.
The degree of damage to strawberry plants during the thaw +5 and following temperature decreases to −15°C (2017–2019).

5. Study of the resistance of generative organs of strawberry to spring frosts

During the years of the project, we did not note the decrease in air temperature and on the soil surface (spring frosts) during the flowering period of strawberry varieties of different ecological and geographical origins. So, the assessment of the degree of damage to generative organs was not made in the field. To accelerate evaluation of resistance of generative organs at horticultural crops, the method of artificial freezing was used [12, 27].

Research works about resistance of generative organs of strawberry to spring frosts by artificial freezing are almost nonexistent currently [28]. During the years of research, experiments on artificial freezing allowed us to conclude that the studied varieties of strawberry are highly resistant to temperatures of −1.0 and – 2.0°C, because visible damages to generative organs were not noted (**Figure 4**).

Decrease of temperature to −2.5°C showed different degrees of damage to the generative organs. In 2018, the percent of dead flowers varied from 0.0 to 70.0% at strawberry varieties. In 2019, the presence of dead flowers was between from 0.0 and 54.6%. On average, for 2 years, dispersion analysis showed significant differences between the studied samples on the percent of flowers damage at 5% significance level. In some varieties ("Kokinskaya rannyaya," "Rosinka," "Tsaritsa"), the flowers could withstand freezing at a temperature of −2.5°C without damage. The Holland variety, "Korona", showed a small amount (5.6%) with damaged pistils. At the control variety, "Urozhainaya TzGL", the flowers were damaged to 16.4%. We noted varieties, on which the flowers were damaged from 33.8 to 46.6% - "Solovushka", "Marmolada", "Sara", "Sonata". At a temperature of −2.5°C, the highest number of damaged flowers was shown—56.3% at the Italian variety "Alba" (**Table 12**, **Figure 5**).

During the years of the research, the buds at varieties of strawberry were damaged less than flowers after an exposure to a temperature of −2.5°C. On average, for 2 years, the buds at some varieties ("Kokinskaya rannyaya," "Rosinka," "Tsaritsa") withstood

"Rosinka" "Tsaritsa"

Figure 4.
Alive flowers and buds of varieties of strawberry after exposure to the temperature of −2.0°C.

Varieties	2018	2019	Average value	Angle-arc sine √percent
	% dead flowers			
"Urozhainaya TzGL" (st)	0.0	32.8	16.4	23.9
"Kokinskaya rannyaya"	0.0	0.0	0.0	0.0
"Solovushka"	0.0	0.0	0.0	0.0
"Rosinka"	38.6	54.6	46.6	43.1*
"Tsaritsa"	0.0	0.0	0.0	0.0
"Alba"	70.0	42.6	56.3	48.7*
"Marmolada"	0.0	11.2	5.6	13.7
"Korona"	33.4	34.2	33.8	35.6*
"Sara"	37.1	30.5	33.8	35.6
"Sonata"	34.8	45.3	40.1	39.3*
$LSD_{0.05}$				32.4

Significantly at a significance level of $p < 0.05$.

Table 12.
The percent of dead flowers of strawberry varieties after exposure to the temperature − 2.5°C, %.

freezing at a temperature of −2.5°C without damage. "Korona," "Sara," and "Sonata" varieties showed a small amount of buds (no more than 10.0%) with damaged pistils. The varieties with damaged buds from 10.5 to 20.1% were noted—"Solovushka," "Urozhainaya TzGL," and "Marmolada". At a temperature of −2.5°C, the highest number of damaged buds was shown—40.8% at the Italian variety "Alba" (**Table 13**).

Exposure to a temperature of −3.0°C increased the damage of flowers in the studied varieties of strawberry. At the same time, the significant intervarietal differences were found in sign of the degree of flower damage at the 5% significance level. A smaller percent of flower damage (6.8%) was noted at "Tsaritsa" variety. It should be noted that the flowers damaged slightly at a temperature of −3.0°C in varieties of "Kokinskaya rannyaya" and "Rosinka"—no more than 18.0%. From the data provided, it can be seen that the flowers were damaged strongly from 53.0 to 73.7% at the varieties "Solovushka," "Urozhainaya TzGL," "Alba," "Korona," "Marmolada," and "Sara." The highest percent of dead flowers after exposure to a temperature of −3.0°C was noted at the Holland variety "Sonata"—84.8% (**Table 14**, **Figure 6**).

a b

Figure 5.
The damaged flowers at "Solovushka" variety (a) and alive flowers at the "Korona" (b) variety after exposure to a temperature of −2.5°C.

Varieties	2018	2019	Average value	Angle-arc sine √percent
	% no dead buds			
"Urozhainaya TzGL" (st)	0.0	21.0	10.5	18.9
"Kokinskaya rannyaya"	0.0	0.0	0.0	0.0*
"Solovushka"	0.0	0.0	0.0	0.0*
"Rosinka"	13.8	22.8	18.3	25.3
"Tsaritsa"	0.0	0.0	0.0	0.0*
"Alba"	61.3	20.3	40.8	39.7
"Marmolada"	0.0	5.8	2.9	9.8
"Korona"	18.0	22.2	20.1	26.6
"Sara"	4.7	8,.5	6.6	14.9
"Sonata"	0.0	14.6	7.3	15.7
$LSD_{0.05}$				20.3

Significantly at a significance level of $p < 0.05$.

Table 13.
The percent of dead buds of strawberry varieties after exposure to the temperature − 2.5°C, %.

After exposure to a temperature of −3.0°C, the buds were not damaged at "Kokinskaya rannyaya," "Rosinka," and "Tsaritsa" varieties. As we can see from **Table 15**, the buds were damaged in the range from 27.7 to 40.1% at "Solovushka," "Urozhainaya TzGL," "Korona," "Marmolada," and "Sara" varieties. The highest percent of dead buds after exposure to a temperature of −3.0°C was noted at the Italian variety "Alba" (50.4%) and the Holland variety "Sonata"—60.6%. Statistical processing of the results of artificial freezing allowed us to determine significant intervarietal differences in the degree of bud's damage at the 5% significance level (**Table 15**).

The temperature regime of −4.0°C was critical for the flowers of the studied varieties of strawberry, which were damaged from 72.3 to 100.0%. Damage of buds was noted from 43.8 to 69.1%.

Varieties	2018	2019	Average value	Angle-arc sine √percent
	% dead flowers			
"Urozhainaya TzGL" (st)	57.6	59.2	58.4	49.8
"Kokinskaya rannyaya"	0.0	25.0	12.5	20.7*
"Solovushka"	6.3	28.8	17.6	24.8*
"Rosinka"	80.5	66.8	73.7	59.2
"Tsaritsa"	0.0	13.6	6.8	15.1*
"Alba"	79.8	26.8	53.3	46.9
"Marmolada"	63.3	79.7	71.5	57.7
"Korona"	47.0	59.0	53.0	46.9
"Sara"	73.8	40.4	57.1	49.1
"Sonata"	100.0	69.5	84.8	67.0
$LSD_{0.05}$				28.2

Significantly at a significance level of $p < 0.05$.

Table 14.
The percent of dead flowers of strawberry varieties after exposure to the temperature – 3.0°C, % (2018–2019).

a

b

Figure 6.
Alive generative organs at the "Tsaritsa" variety (a) and dead at the "sonata" (b) variety after exposure to the temperature – 3.0°C.

Correlation analysis determined a high dependence of the damage degree of flowers (r = 0.97) and buds of strawberries (r = 0.98) on the intensity of the temperature of spring frosts.

Based on the research results, we offer to share the varieties of strawberries by their resistance to spring frosts into five groups:

1. Highly resistant varieties—the number of damaged flowers and buds after freezing at −3.0°C does not exceed to 25.0% and at −2.5°, C10.0%.

2. Resistant varieties—the number of damaged flowers and buds after freezing at −3.0°C is from 25.0 to 50.0% and at −2.5°C is 25.0%.

3. Medium-resistant varieties—the number of damaged flowers and buds at −3.0°C is from 50.0 to 75.0% and at −2.5°C is from 25.0 to 50.0%.

Varieties	2018	2019	Average value	Angle-arc sine $\sqrt{percent}$
	% dead buds			
"Urozhainaya TzGL" (st)	34.4	41.4	37.9	38.0
"Kokinskaya rannyaya"	0.0	0.0	0.0	0.0*
"Solovushka"	0.0	0.0	0.0	0.0*
"Rosinka"	37.8	42.4	40.1	39.3
"Tsaritsa"	0.0	0.0	0.0	0.0*
"Alba"	66.7	34.1	50.4	45.2
"Marmolada"	14.6	55.8	35.2	36.4
"Korona"	17.4	55.0	36.2	37.0
"Sara"	34.1	21.3	27.7	31.8
"Sonata"	90.1	33.1	61.6	51.7
$LSD_{0.05}$				32.1

*Significantly at a significance level of $p < 0.05$.

Table 15.
The percent of dead buds of strawberry varieties after exposure to the temperature – 3.0°C, % (2018–2019).

4. Weakly resistant varieties—damaged flowers and buds after freezing at −3.0°C are more than 75.0% and at −2.5°C from 50.0 to 70.0%.

5. Unstable varieties—the number of damaged flowers and buds at −3.0°C is 100.0% and at −2.5°C more than 75.0%.

So, according to the results of ranking varieties by groups of resistance to spring frosts, we recommend two regimes. The first temperature regime is –2.5°C, which will allow to make mass rejection of unstable forms. The second regime is −3.0°C; it will make possible to select forms that are sources of high resistance to spring frosts for selection for a given sign. The high potential of resistance to spring frosts was shown by varieties of strawberries—"Kokinskaya rannyaya," "Rosinka," and "Tsaritsa."

6. Net productivity of photosynthesis, respiration intensity, potential productivity, growth rate, and yield strawberry varieties

The main elements of the production process of plants are photosynthesis and respiration. In this regard, a pigment analysis, analysis of the photochemical activity of isolated chloroplasts, the intensity of respiration, and the net productivity of photosynthesis were performed.

The efficiency of photosynthetic activity of plants depends from the development of the pigment system. Studies, which were made in leaves of strawberry plants, showed that the "Alba," "Marmolada," "Korona," and "Kokinskaya rannyaya" varieties did not differ significantly from each other in the content of chlorophyll in the leaves and the amount of green pigment was significantly lower in them than in the varieties of "Sara," "Sonata," "Solovushka," and "Tsaritsa" (**Table 16**). A similar regularity in the content of chlorophyll remained in 2019 (**Table 17**). Two other varieties were added; from them the variety "Rosinka" was referred in the group with a low content of green pigment and "Urozhainaya TzGL" in the group with high content. Higher level of chlorophyll at the "Sara," "Sonata," "Solovushka," "Tsaritsa," and "Urozhainaya TzGL" varieties is associated probably with an increased content of other carotenoid pigments in the photosynthetic apparatus. Thus, the correlation

Varieties	Pigments, mg/g		PhCA, microMol $K_3[Fe(CN)_6]$/(mg chl· h)	NPP, g/(m^2 day)	respiration intensity, ml CO_2/g·h
	Chlorophyll	Carotenoids			
"Kokinskaya rannyaya"	2.10 ± 015	0.18 ± 0.006	16.16 ± 0.45	1.38 ± 0.06	3.00 ± 0.39
"Solovushka"	2.33 ± 0.02	0.19 ± 0.010	11.19 ± 0.42	10.25 ± 0.57	3.22 ± 0.60
"Tsaritsa"	2.36 ± 0.03	0.20 ± 0.010	12.71 ± 0.70	8.53 ± 0.35	3.06 ± 0.67
"Alba"	1.98 ± 0.02	0.16 ± 0.006	6.30 ± 0.58	9.37 ± 0.44	2.85 ± 0.20
"Korona"	2.10 ± 0.10	0.18 ± 0.006	14.27 ± 0.66	0.77 ± 0.09	2.87 ± 0.57
"Marmolada"	2.02 ± 0.04	0.16 ± 0.006	7.77 ± 0.26	2.03 ± 0.24	3.61 ± 0.33
"Sara"	2.52 ± 0.02	0.20 ± 0.006	2.48 ± 0.26	3.65 ± 0.32	3.52 ± 0.44
"Sonata"	2.32 ± 0.01	0.20 ± 0.006	12.60 ± 0.66	7.97 ± 0.35	3.13 ± 0.70
$LSD_{0.05}$	0.19	0.02	1.63	0.90	0.65

Table 16.
Indicators of photosynthetic activity and respiration rate of strawberry plants in 2018.

Varieties	Pigments, mg/g		PhCA, microMol $K_3[Fe(CN)_6]$/(mg chl· h)	NPP, g/(m^2 day)	respiration intensity, ml CO_2/g·h
	Chlorophyll	Carotenoids			
"Urozhainaya TzGL"	2.15 ± 0.03	0.17 ± 0.003	8.65 ± 0.26	3.12 ± 0.36	5.03 ± 0.39
"Kokinskaya rannyaya"	2.12 ± 0.02	0.18 ± 0.002	12.88 ± 0.52	3.34 ± 0.37	5.41 ± 0.24
"Solovushka"	2.35 ± 0.03	0.20 ± 0.003	10.80 ± 0.40	8.90 ± 0.64	6.65 ± 0.26
"Rosinka"	2.25 ± 0.03	0.18 ± 0.003	11.50 ± 0.46	2.47 ± 0.27	4.30 ± 0.29
"Tsaritsa"	2.50 ± 0.02	0.19 ± 0.002	8.50 ± 0.29	9.50 ± 0.40	3.21 ± 0.28
"Alba"	2.20 ± 0.05	0.16 ± 0.003	4.49 ± 0.31	5.78 ± 0.39	6.87 ± 0.50
"Marmolada"	2.20 ± 0.03	0.17 ± 0.003	10.06 ± 1.00	2.88 ± 0.18	7.57 ± 0.33
"Korona"	2.21 ± 0.04	0.15 ± 0.003	3.53 ± 0.33	3.40 ± 0.29	4.60 ± 0.29
"Sara"	2.57 ± 0.06	0.19 ± 0.003	4.28 ± 0.53	2.66 ± 0.14	2.30 ± 0.29
"Sonata"	2.37 ± 0.02	0.19 ± 0.004	11.93 ± 0.54	7.33 ± 0.19	3.83 ± 0.33
$LSD_{0.05}$	0.09	0.01	1.44	0.74	1.02

Table 17.
Indicators of photosynthetic activity and respiration rate of strawberry plants in 2019.

coefficient between chlorophyll and carotenoids was 0.92 in 2018 and in 2019, 0.64. As a rule, carotenoids do not only have a light-absorbing function but also protect chlorophyll from photo-oxidation on the background of high solar insolation [29].

Therefore, the observed decrease of the correlation coefficient between chlorophyll and carotenoids in 2019 can be associated with a lower part of solar insolation, which is indirectly evidenced by cooler daytime conditions of the growing season (**Table 17**). As a result, the level of chlorophyll was less dependent from the antioxidant properties of carotenoids. However, the increased content of carotenoids was noted consistently in the group of varieties with high level of chlorophyll during 2 years of research.

To determine the potential abilities of the photosynthetic apparatus, we used the characteristic of the functional activity of chloroplasts on the level of light reactions [30]. As a result, it was shown that the highest speed of Hill reaction the varieties

"Korona", "Sonata", "Solovushka", "Tsaritsa", "Kokinskaya rannyaya", "Rosinka", and "Urozhainaya TzGL" had. At the same time, it should be noted that the correlation between the amount of green pigment and the speed of energy transfer in photosystems was quite low and had a negative value, during the 2 years of the study (in 2018, r = −0.23, in 2019, r = −0.22). So, at the "Sara" variety, despite high indicators of chlorophyll in the light-collecting complex of the leave, the speed of energy transfer in photosystems was low. At the same time, at the varieties "Urozhainaya TzGL", "Rosinka," and "Kokinskaya rannyaya" was a high speed of light reactions on the background of a low level of green pigment. The lack of connection between the amount of green pigment and the PhCA of isolated chloroplasts can be associated with the degree of strength of the chlorophyll-lipoprotein bond [31, 32]. According to research by Zakarian N.E. et al. [33], it was shown that the photochemical activity of isolated chloroplasts *Solanum tuberosum*, *Chrysanthemum indicum*, Trifolium pretense, Pelargonium, and Hibiscusrosa cninensis sharply decreased by 5.4–13.0 times in the lack of a labile form of chlorophyll compared to control plants where both forms of green pigment were present.

When determining the net productivity of photosynthesis (NPP) (as an integral indicator of photosynthetic activity of plants), it was shown that only the "Alba," "Sonata," "Solovushka," and "Tsaritsa" varieties were characterized by an effective accumulation of plastic equivalents for 2 years of research. At the same time, the "Korona," "Kokinskaya rannyaya," and "Urozhainaya TzGL" varieties had a low level of NPP on the background of a low content of chlorophyll and an increased rate of electron transfer to the light phase of photosynthesis. On the contrary, the Italian variety "Alba," which had a low speed of Hill reaction, was characterized by intensive assimilation of carbon dioxide and was at the level of indicators of the varieties "Sonata," "Solovushka," and "Tsaritsa" in 2018, which had an increased content of chlorophyll and had a high efficiency of NPP of chloroplasts.

In addition of research the features of photosynthetic activity, it was important to establish the effectiveness of the distribution of assimilates, namely, to consider the donor-acceptor relationship between maturing fruits and the leave's apparatus. Indirectly, we can discuss about the outflow of assimilates by studying the ratio of dry mass of forming berries to leaves in dynamics [34]. The analysis of this ratio showed that at the varieties "Alba," "Sonata," "Solovushka," and "Tsaritsa" in 2018, this indicator, on contrast to other varieties, increased most intensively in the direction of forming berries, which could indicate about the more active outflow of plastic substances into the maturing berries (**Figure 7a**). However, in 2019, at the varieties of "Solovushka" and "Sonata," the intensity of the outflow of assimilates into maturing fruits was significantly decreased compared to the first year of the study, and in "Sara," on the contrary, it was increased. Really, at "Solovushka" and "Sonata" varieties, there was a great accumulation of dry substance of leaves. So, the "Solovushka" and "Sonata" varieties increased the dry substance at 1.78–2.69 times, compared to the first decade of June to the second decade of July, and the "Sara" variety did not increase it significantly. In other varieties, the increase in dry leave's biomass in the second decade of July compared to the first decade of June was at 1.15–1.40 times. At the same time, the "Alba," "Korona," and "Tsaritsa" varieties retained an active outflow of plastic equivalents to the ripening berries in 2019.

In addition, the analysis of growth activity, which was evaluated by the speed of accumulation of raw plant substance, showed that the varieties "Alba," "Sonata," "Solovushka," and "Tsaritsa" had differences in this indicator significantly in 2018. The activity of accumulation of raw plant substance varied from 8.68 to 9.76% in these varieties (against the others from 2.41 to 5.43%) (**Figure 7b**). In 2019, as in 2018, the relative stability and high speed of accumulation of raw substance of the whole plant, showed the "Sonata" and "Tsaritsa" varieties. In 2019, the "Alba," "Sara," and "Solovushka" varieties decreased the intensity of accumulation of raw biomass of the whole plant significantly at 2.20–2.65 times. On the contrary, "Kokinskaya rannyaya," "Marmolada," and "Korona" increased it in 2019 compared to 2018.

Determining the respiratory intensity, as one of the main components of the production process in 2018 did not reveal significant differences between the studied varieties. However, in 2019, it was higher temperature conditions of the growing season at night, and strawberry varieties showed significant differences of the respiratory intensity. As a result, in 2019, there was a probability of excessive waste of plastic substances in strawberry plants that were formed as a result of photosynthesis and participate in the production process. So, the "Alba," "Solovushka," "Korona," "Kokinskaya rannyaya," and "Rosinka" varieties were characterized by a more intense breathing. At these varieties, the release of carbon dioxide varied from 5.0 to 7.57 ml of CO_2/g·h, while in other varieties from 2.30 to 4.60 ml of CO_2/g·h.

In the future, it was interesting to determine how the identified photosynthetic features of the studied varieties could affect the production process. Sometimes there could be either no direct dependent between the intensity of photosynthesis and productivity, or there could be a negative correlation between these indicators [35].

The result of the production process is evaluated by the yield or the share of useful product in the total mass of the plant. It is shown that the "Alba," "Sonata," "Solovushka," and "Tsaritsa" varieties had the highest indicators of berry yield on the background of more efficient photosynthetic activity and increased outflow of assimilates in 2018 (**Figure 8**). The varieties of domestic selection—"Tsaritsa" and "Solovushka"—were shown especially to have high yield. However, in 2019, only the "Korona" and "Sonata" produced consistently high yields. In 2019, the varieties "Alba," "Solovushka," and "Korona" decreased significantly the yield of berries at

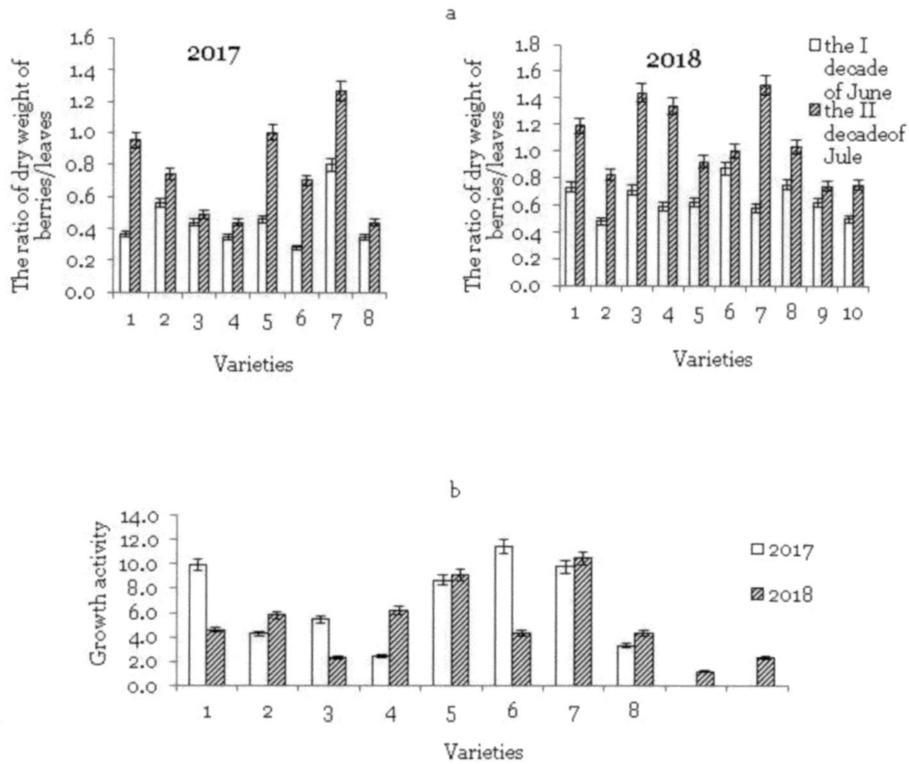

Figure 7.
The ratio of dry weight of berries/leaves (a) and the activity of accumulation of the plant raw biomass (b). Varieties: (1) "Alba," (2) "Marmolada," (3) "Sara," (4) "Korona," (5) "sonata," (6) "Solovushka," (7) "Tsaritsa," (8) "Kokinskaya rannyaya," (9) "Urozhainaya TzGL," and (10) "Rosinka".

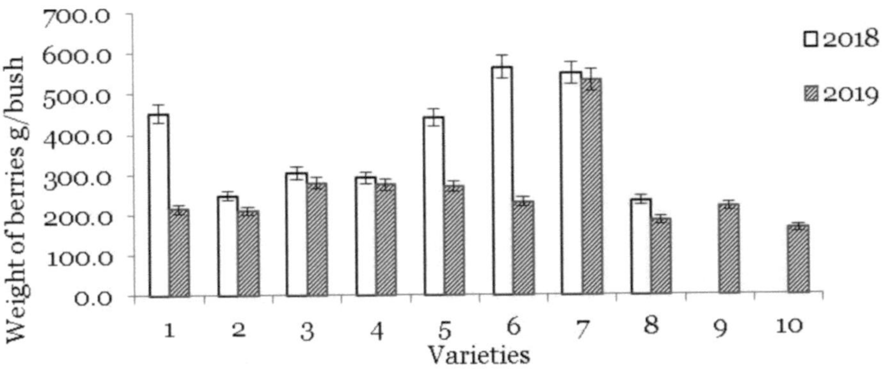

Figure 8.
The yield of the varieties of strawberry. Varieties: (1) "Alba," (2) "Marmolada," (3) "Sara," (4) "Korona," (5) "sonata," (6) "Solovushka," (7) "Tsaritsa," (8) "Kokinskaya rannyaya," (9) "Urozhainaya TzGL," (10) "Rosinka."

1.64–2.40 times as a result of both increased respiratory intensity and violations of donor-acceptor relations. The other varieties—"Marmolada," "Sara," "Kokinskaya rannyaya," "Urozhainaya TzGL," and "Rosinka"—had a low yield index to compare with "Tsaritsa" [18].

So, the research conducted during 2 years showed that only the one variety of strawberry—"Tsaritsa"—in the conditions of the middle zone of Russia had an effective work on the photosynthetic apparatus and increased outflow and accumulation of assimilates in the maturing fruit, which, as a result, affected its high yield. Other varieties showed or not stable yield over the years, such as "Solovushka," "Alba," and "Sonata" or low yield as in "Marmolada," "Sara," "Korona," "Kokinskaya rannyaya," "Urozhainaya TzGL," and "Rosinka." In these cultivars, the reasons of low yield were increased respiratory intensity or insufficient efficiency of the photosynthetic apparatus and the violations of donor-acceptor relations.

At the same time, we found a very high degree of dependence between the actual yield of strawberry varieties and the net productivity of photosynthesis ($r = 0.88$). At the same time, a moderate dependence was found between the degree of freezing of strawberry plants in winter and the actual yield on the one hand ($r = -0.47$) and between the net productivity of photosynthesis ($r = -0.36$) on the other. The calculation of the triple correlation coefficient between the actual yield, the net productivity of photosynthesis, and the degree of freezing of plants in winter showed a very high dependence ($r = 0.89$) between the studied indicators.

7. Conclusion

We had studies of physiological and biochemical processes before and after autumn hardening of strawberry varieties. We studied varieties of strawberries, based on physiological and biochemical changes, that characterized the state of plants after exposure to low positive and negative temperatures at the beginning of the winter period. We evaluated the damage of the rhizome tissues and the branch crowns of strawberry varieties of different ecological and geographical origin in the early winter period and during thaws at the winter after artificial freezing and in the field. We evaluated the damages of generative organs by method of artificial freezing during the flowering period of strawberry, and we offered a ranking for groups of resistance to spring frosts. We conducted comparative physiological studies of the photosynthetic apparatus of different varieties of strawberries of different ecological and geographical origin by productivity.

As a result of the research, it was found that in the autumn and early winter, the increase in bound water and the decrease in free water in the leaves were characteristics for strawberry plants on the background of a decrease at the level of hydration. The change in the composition of water fractions was dependent more on the accumulation of sucrose and free proline in the leaves of plants during the autumn period. At the same time, "Solovushka," "Tsaritsa," "Sara," and "Korona" had the highest bound water/free ratio by the end of autumn and beginning of winter, less damage in the structural and functional integrity of cell membranes, a low level of accumulation of hydrogen peroxide, and an intensification of the activity of the antioxidant enzyme catalase. All this indicates that these varieties are characterized by high adaptive ability in the climatic conditions of Central Russia.

As a result of artificial freezing, it was noted that the temperature decrease to −15°C at the end of November did not cause of irreversible damage at strawberry varieties. The decrease of temperature in early December to −20°C increased the damage at the studied varieties of strawberries. During the winter, it was noted that the reaction of strawberries was increasing to the thaw by the end of the winter period, which was associated with the resumption of growth processes after the influence of positive temperatures. As a result of the research, frost-resistant varieties were identified; these were "Solovushka," "Tsaritsa" (Russia), and "Korona" (Holland).

As a result of the damaging factors of the spring period, high ability of resistance to spring frost, showed the varieties of strawberry—"Kokinskaya rannyaya," "Rosinka," and "Tsaritsa," According to the results of varieties ranking by groups of resistance to spring frosts, we can recommend two regimes. The first temperature regime is −2.5°C, which will allow mass rejection of unstable forms. The second regime −3.0°C makes it possible to select forms that are sources of high resistance to spring frosts for selection for a given sign.

Researches on determine the net productivity of photosynthesis, respiration intensity, potential productivity, growth rates and yield of strawberry varieties of different ecological and geographical origin showed that the "Tsaritsa" variety in the Central region of Russia had an effective work of the photosynthetic apparatus and the outflow of assimilates into the maturing fruit, which, as a result, affected its high yield. On the results of identifying the characteristics of functional connections between physiological and biochemical processes of resistance to the action of a low-temperature environment factors and productivity of strawberry varieties of different ecological-geographical origin, we identified a variety of domestic breeding "Tsaritsa" for creation of resistance berries' agrobiocenosis.

Author details

Zoya Evgenievna Ozherelieva*, Pavel Sergeevich Prudnikov, Diana Aleksandrovna Krivushina, Marina Ivanovna Zubkova and Anna Androsova
Russian Research Institute of Fruit Crop Breeding, Orel, Russia

*Address all correspondence to: ozherelieva@vniispk.ru

References

[1] Aaby K, Ekeberg D, Skrede G. Characterization of phenolic compounds in strawberry fruits by different HPLC detectors and contribution of individual compounds to total antioxidant capacity. Journal of Agricultural and Food Chemistry. 2007;**55**:4395-4406

[2] Aitzhanova SD. Selection of strawberries in the South-Western part of the non-chernozem zone of Russia [abstract PhD dissertation in agricultural sciences]. Bryansk; 2002

[3] Eremin GV, Efimova IL. Rootstocks of Seed and Stone Crops for Modern Intensive Industrial Technologies. Developments, that Form the Modern Image of Gardening. Krasnodar: GNU SKZNIISiV; 2011. pp. 118-139

[4] Mukhanin IV, Zhbanova O. Modern assortment of strawberry for industrial cultivation. Russian school of horticulture. Strawberry. 2015;**1**:19-24

[5] Efimova IL, Shaforostova NK, Kuznetsova AP. Adaptive and productive potential of rootstocks of fruit crops in the conditions of southern gardening. Fruit and Berry Growing in Russia. 2008;**XVIII**:135-141

[6] Krasova NG, Ozhereleva ZE, Golyshkina LV, Makarkina MA, Galasheva AM. Winter Hardiness of Apple Varieties. Orel: VNIISPK; 2014. p. 184

[7] Abyzov VV. Biological features and economic value of strawberry varieties in the conditions of Central Russia [dissertation]. Michurinsk; 2008

[8] Govorov VN. Assessment of the Stability of New Varieties and Hybrids of Strawberries to the Main Fungal Diseases and Pests in the Central Zone of the Krasnodar Territory. Krasnodar; 2011. p. 22

[9] Govorova GF, Govorov DN. Strawberry: Past, Present, Future. Moscow; 2004. p. 347

[10] Shokayeva DB. Biological Bases and Regularities of Strawberry Fruiting. Orel: Kartush; 2006. p. 183

[11] Stolnikova NP. Culture of Strawberry in Western Siberia. GNU NIISS. Barnaul: IP Kolmogorov I.A; 2014. p. 182

[12] Ozhereleva ZE, Zubkova MI, Krivushina DA. Accelerated evaluation of resistance to spring frosts of strawberry varieties. Fruit and Berry Growing in Russia. 2018;**54**:123-126. DOI: 10.31676/2073-4948-2018-54-123-126

[13] Makarova NV, Stryukova AD, Antipenko MI. Comparative analysis of the chemical composition and antioxidant activity of repair and non-repair varieties of strawberry. Storage and Processing of Agricultural Raw Materials. 2014;**9**:45-48

[14] Tyurina MM, Gogoleva GA, Goloulina LK, Morozova NG, Echedi YY, Volkov FA, et al. Determination of Stability of Fruit and Berry Crops to Cold Season Stressors in Field and Controlled Conditions. Moscow; 2002. p. 120

[15] Yakushkina NI, Bakhtenko EY. Physiology of Plants. Moscow: Vlados; 2004. p. 464

[16] Matala V. The Cultivation of Strawberries. Saint Petersburg; 2003. p. 202

[17] Ozherelieva ZE, Prudnikov PS, Bogomolova NI. Frost hardiness of introduced *Hippophae rhamnoides* L. genotypes in conditions of Central Russia. Proceedings of the Latvian Academy of Sciences, Section B: Natural, Exact and Applied Sciences. 2016;**70.2**(701):88-95. DOI: 10.1515/prolas-2016-0014

[18] Prudnikov PS, Ozhereleva ZE, Krivushina DA, Zubkova MI. The formation of resistance to hyperthermia of FRAGARIA x ANANASSA DUCH varieties of different ecological and geographical origins in the autumn period. Vegetables of Russia. 2019;**6**:80-83

[19] Dzhavadian N, Karimzade G, Mafuzi S, Ganati F. Changes in the activity of enzymes and the content of proline, carbohydrates and chlorophyll in wheat caused by cold. Physiology of Plants. 2010;**57**(4):580-588

[20] Prudnikov SP, Gulyaeva AA. Features of hyperthermia effect on the hormonal system and antioxidant status of *Prunus armeniaca* L. Breeding and Selection Horticultural Crops. 2015;**3**:151-154

[21] Prudnikov PS, Krivushina DA, Gulyaeva AA. Components of the antioxidant system and intensity of POL in *Prunus cerasus* L. under the action of hyperthermia and drought. Breeding and Selection Horticultural Crops. 2016;**3**:116-119

[22] Luo Y, Tang H, Zhang Y. Production of reactive oxygen species and antioxidant metabolism about strawberry leaves to low temperatures. The Journal of Agricultural Science. 2011;**3**:89-96

[23] Chudinova LA, Orlova NV. Physiology of Plant Resistance: A Textbook for a Special Course. Perm: Perm University; 2006. p. 124

[24] Shokayeva D.B., Zubov A.A. Strawberry/ program and method of variety study of fruit, berry and nut crops. by E.N. Sedova, T.P. Ogoltsova. Orel: VNIISPK; 1999. pp. 417-443

[25] Lindén L, Palonen P, Hytönen T. Evalution of three methods to assess winter-hardiness of strawberry genotypes. Acta Horticulturae. 2002;**567**:325-328. DOI: 10.17660/ActaHortic.2002.567.69

[26] Ozhereleva ZE, Prudnikov PS, ZubkovaMI, KrivushinaDA, KnyazevSD. Determination of Frost Resistance of Strawberries under Controlled Conditions (Methodological Recommendations). Orel: VNIISPK; 2019. p. 25

[27] Ozhereleva ZE, Zubkova MI, Krivushina DA. Assessment of the resistance of strawberries to spring frosts. Problems of Agricultural Development in the Region. 2019;**4**(40):113-118

[28] Leonchenko VG, Evseeva RP, Zhbanova EV, Cherenkova TA. Previous Selection of Prospective Genotypes of Fruit Plants for Ecological Stability and Biochemical Value of Fruits (Methodological Recommendations). Michurinsk; 2007. p. 72

[29] Ladygin VG, Shirshikova G. Modern concepts of the functional role of carotenoids in eukaryotic chloroplasts. Journal of General Biology. 2006;**3**:163-190

[30] Gavrilenko VF, Zhigalova T. Big Workshop on Photosynthesis. Moscow: Akademiya; 2003. p. 256

[31] Sudina EG. State of Chlorophyll and its Photochemical Activity. I Union Biochemical Congress. Publishing House of the USSR Academy of Sciences; 1963. p. II

[32] Sudina EG, Lazovaya GI. The significance of the molecular organization of pigment-containing complexes for functional activity. In: Materials of the II Union Symposium on the Application of Electron Microscopy in Botanical Research. Kiev; 1967. pp. 11-114

[33] Zakarian NE, Tsovyan Zh V, Virabyan AR. About the significance

of various forms of chlorophyll in the process of photosynthesis. Scientific notes of Yerevan state university. Natural science. Biology. 1987;**2**:137-142

[34] Bobodzhanova MD. Photosynthesis, donor-scceptor relations and productivity of medium-fiber cotton (*Gossypium hirsutum* L.) [PhD dissertation in agricultural sciences]. Dushanbe; 2007

[35] Abdulaev Kh A, Giiasidinov BB, Solieva BA, Mirakilov Kh M. Paradox: photosynthesis of cotton, as a conservative feature, has not changed significantly for a hundred years of selection of new varieties. In: Materials of the Russian Scientific Conference with International Participation Plants in the Conditions of Global and Local Natural-Climatic and Anthropogenic Impacts. Petrozavodsk; 2015. p. 19

Chapter 4

Castor (*Ricinus communis*): An Underutilized Oil Crop in the South East Asia

Swapan Chakrabarty,
Abul Kalam Mohammad Aminul Islam, Zahira Yaakob and Abul Kalam Mohammad Mominul Islam

Abstract

Castor belongs to a monotypic genus *Ricinus* and subtribe Ricininae. It is one of the oldest plants, getting importance as an agricultural crop for subtropical and tropical countries in the world. Castor is a hardy plant, requires low input, tolerates marginal soils, is easy to establish in the field, is resistant to drought, and gives yield 350–900 kg oil per hectare. Castor oil shows great functional value in energy sector, industry, and pharmaceutical. In recent years, it received increasing demand in the international market for its more than 700 uses, ranging from medicine and cosmetics to biodiesel, plastic, and lubricants. The oil is significant for many industrial uses compared with other oils from plant sources because of its high and low temperature-tolerant properties. This chapter has been written to provide botanical descriptions, ecology, agro-technology, and versatile industrial uses.

Keywords: *Ricinus communis*, biodiesel, underutilized crops, monotypic

1. Introduction

Castor (*Ricinus communis* L.) is a nonedible vegetable oil seed crop cultivated all over the world. It grows well in the wet tropics to the subtropical dry region with an optimum temperature of 20–25°C [1]. Though Ethiopia is considered as the probable center of origin, its cultivation is mostly concentrated in Asia, about 92.2% of the total production [2]. India is the largest producer and exporter of the castor oil (www.statista.com). Castor oil is an important feedstock for the chemical industry which is mostly utilized for biodiesel production and pharmaceutical application because it is the only commercial source of hydroxylated fatty acid [3]. It is also used for the production of lubricants, hydraulic and brake fluid, polymer materials, coating, fertilizer, soaps, waxes, and greases. The use of castor oil in Southeast Asian countries is limited due to its high cast of detoxification of castor meal and refining of castor oil. The plant also contains toxic compounds such as ricin, ricinin, and RCA (*Ricinus communis* agglutinin) [4]. The major challenges for castor production are the development of high-yielding varieties that will be non-shattering, dwarf, resistance or tolerance to disease and insect pests, and low ricin, ricinin, and RCA content. The development of easy method for the detoxification of castor meal and refining of castor oil is also important. Castor oil has a great future potential for use in the production of biodiesel.

2. Castor plant

Castor (*Ricinus communis* L.) is a species of annual or perennial flowering plant which belongs to the spurge family Euphorbiaceae, monotypic genus *Ricinus*, and Ricininae subtribe. It is a fast-growing and suckering shrubby tree which reaches the size of 5–12 m. It is commonly known as veranda (Bengali), arandi (Hindi), era-gach (Assamese), castor and castor oil plant (English), wonderboom (Dutch), ricin (French); Rizinus and Palma Christi (German), Fico d'inferno (Italy), and ricino (Portuguese).

3. Origin and distribution

Castor is generally distributed in tropical, subtropical, and warm-temperate regions of the world. It is very commonly found in fellow land, roadside, and compounds in rural and urban areas and also common along seasonally dry rivers in altitudes between 400 and 2700 m. The probable center of origin for castor is Northeastern Africa, i.e., Ethiopia and Somalia [5], and it has four centers of diversity, viz., (a) Ethiopian-Eastern African, (b) Northwest and Southwest Asia and Arabian peninsula (c), subcontinent of India, and (d) China [3]. It is currently naturalized across the African continent, the Atlantic coast to the Red Sea, Tunisia to South Africa, and islands in the Indian Ocean. It is also widely cultivated and naturalized in tropical and subtropical regions of America and Asia and temperate areas of Europe [6].

4. Botany and reproductive biology

4.1 Botany

Leaves: The long-stalked and simple glazing leaves are 15–45 cm long; alternate and palmate with 5–12 deep lobes with serrate leaf margins (**Figure 1**); lobes acuminate, membranous, oblong to linear; 1–3-cm-long stipules united to a sheathing bud, deciduous; and petiole 3.5–50 cm long, round [7].

Different leaf colors are observed in castor, which start off as dark reddish purple or bronze when young and turn into dark green, sometimes with a reddish tinge as they mature. In some varieties, the leaves are really green from the start, whereas in others, a pigment suppresses the green color of all chlorophyll-bearing parts, leaves, stems, and young fruits so that they remain a dramatic purple to reddish brown color throughout the whole life of the plant.

Flower: The flowers are burgeoned in an erect terminal panicle-like inflorescence, which consists of cymes, usually glaucous, later-appearing lateral by overtopping, up to 40 cm long. The flowers are unisexual, regular, with short pedicel, 1–1.5 cm in diameter; the calyx with 3–5 lobes; corolla absent; male flowers toward the base of the inflorescence with many stamens in branched bundles; and female flowers relatively few in number and remain toward the apex of the inflorescence with early caducous sepals, three-celled superior ovary, usually soft spiny, style 3, red or green, 2-cleft [7].

Fruit: Fruits are ellipsoid to subglobose, usually three-lobed smooth or spiny capsule (**Figure 2**), 1.5–2.5 cm long, brown, dehiscing in three cocci each opening by a vulva and one-seeded [8].

Seed: Seed is ellipsoid, 9–17 mm long, compressed with a brittle, mottled, glaring seed coat with distinct caruncle at the base, endosperm copious, white, and cotyledons thin [8].

Seedling: Seedlings are grown by epigeal seed germination; cotyledons petioled, broadly oblong up to 7 cm long, flat with entire margins; and first leaves opposite [8] (**Figures 1** and **2**).

4.2 Reproductive biology

In angiosperms, the flowers are the reproductive structures which are most varied physically and show a correspondingly great diversity in the methods of reproduction. It reproduces by following a mixed pollination system, which prefers selfing by geitonogamy and, at the same, outcrossing by anemophily and entomophily. Under natural condition more than 80% cross-pollination occurs. Flowering may occur within 6 months after seed germination. The flowers of castor are normal monoecious, i.e., it bears pistillate flower on the upper part of the raceme and staminate on the lower part [9]. The proportion of pistillate and staminate flowers among racemes can vary both within and among genotypes and also influenced by the environment. The percentage of pistillate flowers in normal monoecious varieties is highest on the first racemes and decreasing subsequently on developed racemes. The number of staminate flowers is proportionally increased with the decrease in pistillate flowers [10]. The probable cause of this variation is mainly temperature in different season. Moderate temperature in spring and early summer promotes female flower, while high temperature in mid and late summer

Figure 1.
(a) Castor plant, (b) leaf, (c) inflorescence, (d) seedling, (e) root. Source: The above pictures were collected from Habiganj, Sylhet, Bangladesh.

Figure 2.
Castor fruit and seeds, (a) spiny fruit, (b) spineless fruit. Source: the above pictures were collected from Habiganj, Sylhet, Bangladesh.

promotes male flower. Femaleness is highest in young plants with a high level of nutrition, and maleness is highest in older plants with a low level of nutrition [11]. After opening, a male flower releases viable pollen grains for 1–2 days, and shedding of pollen occurs in the morning. The temperature between 26 and 29°C and relative humidity of 60% is the best environmental conditions for pollen dispersal which may vary according to cultivar. Before the opening of the female flowers, male flowers reach its maturity, and anthesis usually occurs in a short period of time [12]. Therefore, the pistillate flowers that open and become receptive get a large source of pollen. After the opening of the flower, the stigma remains fully receptive for few hours, but there is a difficulty for pollination to occur promptly after the opening of the flower. Depending on the environmental condition, the stigma may remain receptive for 5–10 days [13].

5. Climate and soil

Castor, a ruderal species, can be grown in tropical and subtropical regions as a perennial plant and in temperate regions as annual plant. It is a long day plant, and for normal growth and development, a day length of about 12–18 hours is required. It is found from sea level to altitudes of about 2000 m with an optimum of 300–1500 m, and its cultivation is restricted to countries lying between 40°N and 40°S latitudes, but in Russia, a few varieties are grown even up to 52°N latitude [8], and in India it is being cultivated up to an altitude of 2500 m. Its cultivation is restricted up to 500 m where frost is common during cropping season. For obtaining better yield, a frost free-growing period of about 140–190 days is highly essential depending on variety. It is deep rooted and fairly drought-resistant crop which requires annual rainfall of 250–750 mm. A moderate temperature of 20–25°C [1] and low relative humidity with clear sunny days are highly favorable for better yield. For germination, a soil temperature of 12–18°C is suitable. It is sensitive to high relative humidity and temperature above 40°C and below 15°C, which yields a negative impact on yield [14].

Castor can be grown well on fairly deep, moderately fertile, slightly acidic, and well-drained sandy loam soils. Heavy clays with poor drainage and marshy soils are not suitable for castor cultivation. The soil with low water holding capacity, pH > 9 and pH < 4, electrical conductivity (EC) >4 dS/m, and exchangeable sodium percentage (ESP) >20% is not suitable for castor cultivation. Moderately fertile soils are preferable than excessive fertile because excessive fertility favors vegetative growth in the expense of seed yield [14].

6. Cultivation technology

6.1 Land preparation

Land preparation is an essential step to prepare soil favorable for cultivation, to control weed, and to conserve soil moisture. Castor requires moist and well-pulverized loose top soil for better seed germination and early growth. It is done by three to four deep plowings followed by two to three harrowing to break the clods and leveling the field. Ridging is recommended in dry areas where the total rainfall is low [8].

6.2 Seed rate, spacing

To cover one hectare of land, a seed rate of 10–12 kg is recommended, but it varied upon cultivars and sowing method. The seed rate will be 8–10 kg/ha for hand dibbling and for intercropping; it depends on sowing proportion component crops. The spacing of castor plantation varies with growth habit, duration of variety, and sowing time. Early and medium duration cultivars are sown at a closer spacing of 90 cm × 45 cm, and long duration cultivars are sown at a wider spacing of 90 cm × 60 cm under rainfed condition. 90 cm × 60 cm spacing is favorable under irrigated condition. However, under late sown situation, a narrow spacing (60 cm × 60 cm) is practiced to realize higher yields [14].

6.3 Propagation and sowing

Castor is generally propagated through seeds and sown during June and July. However, it can be cultivated year-round under irrigated condition. Under rainfed conditions, the seeds are sown by plow furrow, and the seeds are dibbled by maintaining proper spacing under irrigated condition. The emergence of seedling is easy due to its epigeal seed germination behavior. Deep sowing (8–10 cm) is recommended in light soils under rainfed conditions, and shallow sowing (6–8 cm) is preferable under irrigated condition and heavy soils [14].

6.4 Irrigation

Although castor can tolerate moisture stress, it performs well to irrigation. If castor expertizes moisture stress during seedling to flowering stage, it shows poor performance. If irrigation facilities are available, two to three irrigations should be given during this critical stage. If only one irrigation is available, it should be given at flowering stage. No irrigation should be given during maturity stage because it delayed maturity and also influences new vegetative growth [14]. During the season of high rainfall, proper drainage facilities should be provided to avoid water stagnancy.

6.5 Weed management

Weed management is the most important intercultural operation in any crop cultivation as it impacts on overall yield by competing for nutrients, fertilizer, manure, light, and water. The critical period of crop-weed competition is the 50–60 days after sowing (DAS). After land preparation by deep plowing followed by harrowing, the crop needs two weedings and hoeings either manually or mechanically at 25 and 50 days after sowing. Chemical control like weedicides is also effective in controlling weeds. Preplant incorporation of fluchloralin or pre-emergence application of pendimethalin at 1 kg/ha is an effective control measure of grasses and broad-leaved weeds. One of these weedicide along with one hoeing at 50 days after sowing may be effective in controlling weeds in castor [14].

6.6 Cropping and intercropping

As castor is a tall statured crop, it is being cultivated as shade crop for turmeric and also grown as trap crop for pests. It can be cultivated as sole crop in rotation with wheat and linseed. Groundnut, black gram, green gram, sorghum, pearl millet, cowpea, pigeon pea, and cotton can be grown along with castor. Intercropping of castor with pigeon pea extenuate the occurrence of *Spodoptera litura* [14]. In rainfed condition, intercropping of castor with green gram or black gram in a 1:2 ratio and, in irrigated condition, intercropping of castor with onion in a 1:2 ratio by maintaining 1.5 × 1 m spacing are recommended.

6.7 Manures and fertilizers

The application of manures and fertilizer in appropriate time and dose assured better crop growth and yield. More fertilizers are required for hybrids and irrigated crop than variety and rainfed crop. Before final land preparation 15–20 t/ha well-decomposed farm yard manure (FYM) should be applied in both irrigated and rainfed condition for supplying nutrients as well as for moisture conservation. Under rainfed condition, the recommended dose of both N and P_2O_5 fertilizers is 40 kg/ha. At the time of sowing, all P fertilizer with 50% N fertilizer should be applied, and the remaining N fertilizer should be top dressed after first weeding. Under irrigated condition, both P_2O_5 and K_2O are required at a dose of 40 kg/ha, and N fertilizer may be required at 150 kg/ha. These N fertilizers should be applied at three splits at sowing, first hoeing and preflowering stages. As a source of P fertilizer, single superphosphate is more preferable because Ca and S fertilizers are also applied [14].

6.8 Pest and diseases control

Capsule borer (*Dichocrocis punctiferalis*) and semilooper (*Achaea janata*) can be controlled by dusting BHC 10% in early stages or spraying 0.1% carbaryl on the crop. To protect the crop from seedling blight, water logged and low-lying areas should be avoided. To prevent the occurrence of root rot and *Alternaria* blight, castor seed should be treated with thiram or agrosan GN at 3 g/kg seed. In the latter stage of *Alternaria* blight of castor, foliar application of carbamates or copper-based fungicide may be effective. In preventing castor wilt, treatment with carbendazim + *Trichoderma* at 10 g/kg seed + soil application of *Trichoderma* has been found most effective. Seed treatment with *Trichoderma viride* and FYM and neem cake application has also found effective in decreasing the occurrence of *Alternaria* blight [14].

6.9 Harvesting and threshing

The annual type of castor requires about 4–9 months to mature depending on the variety, and the perennial type may continue bearing for 10–15 years. Improved varieties with non-shattering capsule should be harvested as soon they are fully dry, but shattering type capsules are harvested when the capsule turn greenish to yellowish [8]. The central spike on the main rachis matures first than the spike on side branches start maturing. Therefore, two to three pickings may be needed for harvesting the crop because all the spikes do not mature at the same time. Harvesting of immature capsules should be avoided as it has negative impacts on oil content. After harvesting, the capsule should be sun dried for 4–5 days, and finally threshing and winnowing are done by manually or mechanically [14].

6.10 Yield

The yield of castor may vary from 1 to 3 tons of seed/ha depending on agroclimatic conditions, crop management practices, and the hybrid or variety used [8]. The percentage of oil is 37% and seed cake 63%.

7. Genetics and breeding

Naturally all forms of castor are diploid and its chromosome number is 2n = 2x = 20 [15]. In nature, crossing of castor occurs freely and produce fertile progeny. Commonly 5–50% outcrossing occurs naturally, but in some dwarf cultivars, it may be ranged 90–100%. Male-sterile and female-sterile lines have been also identified which have a great value in breeding improved varieties [7]. Due to highly outcrossing nature, a great phenotypic variation is observed in characters such as stem color, epicuticular wax (bloom wax), plant height, presence of spines in capsules, branching behavior, leaf shape, sex expression, seed color, and response to environmental condition [16, 17]. It is also possible to exploit genetic polymorphism for quantitative traits in breeding programs [18, 19]. Mass selection of castor is effective when the traits under selection are highly heritable. This selection technique performs more effectively with self-fertilization of selected plants to impede cross-pollination and controlled selection technique to minimize environmental variation [20]. It is also an effective technique for increasing the frequency of pistillate castor plants of the NES type [21]. Cultivars developed by mass selection are Kavkazskaya (in the former USSR), IAC-38 and BRS Energia (in Brazil), and Conver and Kansas (in the United States) [22, 23]. Back cross method has been used to transfer monogenic traits such as dwarf plant stature, spineless capsules, stem color, bloom, non-shattering, plant height, and resistant to wilt. Pedigree selection has been used to select high-yielding families and individual plant within the families. Subsequent progeny test for oil content and resistant to *Fusarium* wilt was done by Fernández-Martínez and Velasco [22] and developed wilt resistant cultivar Fioletovaya. Individual plant selection followed by progeny test was used to develop cultivar Guarany in Brazil [24]. In India, several cultivars with tall and late maturation such as HC 1 to HC 8, EB 16 A, EB 31, S-20, Junagadh 1, Punjab castor 1, Rosy, and MC 1 were developed by using this method [25]. Recurrent selection has effectively decreased the plant height in the cultivar Guarani by successive cycle of selection and recombination of selected lines or individual plants [26, 27]. Hybridization involving single, double, or triple crosses is being used to combine desired traits from different sources. The first commercial hybrid of castor was GCH 3 which was developed in India [3]. A number of other castor cultivars exist to which "Hale" and "Lynn" are dwarf cultivars in the United States, mainly used as pollen parent in hybrid production. Other well-known cultivars are "Rica" and "Venda" in France and "T-3," "CS-9," "SKI-7," and "GCH" series of hybrids in India [8].

8. Castor seed

8.1 Oil composition

Castor oil is the mostly used and economically important seed oil in the world. Castor seed comprises about 40–55% oil, and kernel contains 64–71% oil which is the highest among all cultivated oil crops (**Table 1**). Castor oil is a unique vegetable oil due to high ricinoleic acid content (84.2–94%) which is a monounsaturated and 18-carbon fatty acid. Ricinoleic acid is exceptional from other fatty acids because it has a hydroxyl functional group on the 12th carbon that makes it more polar than other fats. The chemical reactive capacity of the alcohol group also approbates chemical derivatization that is almost impossible with other seed oils. Castor oil is a valuable chemical in feedstock due to its ricinoleic acid content and underling a high price than other seed oils. Besides ricinoleic acid content, some other fatty acids are present in castor oil, which are presented in **Table 2**.

Salimon et al. [34] identified five major triacylglycerol in castor seed oil. These were triricinolein (RRR), diricinoleoylstearoylglycerol (RRS), diricinoleoyloleoyl-glycerol (RRO), diricinoleoyllinoleoylglycerol (RRL), and diricinoleoylpalmitoyl-glycerol (RRP). They also first reported the per cent composition of triacylglycerol present in castor seed oil (**Table 3**) by using high performance liquid chromatography (HPLC) with evaporative light scattering detector (ELSD).

Crops	Average seed oil percentage	Oleic acid	Linoleic acid	Linolenic acid	Ricinoleic acid
Nonedible oil crop					
Castor	40–55%	2–6%	1–5%	0.5–1%	85–95%
Jatropha	35–40%	21.8–44.7%	31.49–47.8%	0.2%	—
Pongamia pinnata	30–40%	44.5–71.3%	10.8–24.75%	2.9–6.3%	—
Edible oil crop					
Canola	42%	57.59–61.41%	15.3–22.3%	10.8–13%	—
Linseed	38%	20.6–23.6%	19–22%	41.9–53.1%	—
Sunflower	48%	27–36%	52–67%	—	—
Soybean	18%	27.3–29.7%	43–56%	4.6–11.4%	—
Palm	52%	20.3–24.7%	0.7–1.8%	—	—

Source: Yadava et al. [28]; Islam et al. [29]; Karmee and Chadha, [30]; Shrirame et al. [31]; Kostik et al. [32]; Islam et al. [33].

Table 1.
Oil composition of castor and other vegetable oil.

Name of acids	Carbon number	Average percentage range
Saturated fatty acid	—	1–2.5%
Unsaturated fatty acid	—	97.5–98.3%
Ricinoleic acid	18:1	84.2–95%
Oleic acid	18:1	2.8–5.5%
Linoleic acid	18:2	4.3–7.3%
Linolenic acid	18:3	0.2–0.5%
Stearic acid	18:0	0.9–1.2%
Palmitic acid	16:0	0.7–1.3%
Dihydroxystearic acid	18:0	0.3–0.5%
Others	—	0.2–0.5%

Source: Salimon et al. [34]; Conceição et al. [35]; Gupta et al. [36].

Table 2.
Average fatty acid composition of castor seed oil.

Triacylglycerol	Composition (%)
Triricinolein (RRR)	84.1
Diricinoleoylstearoylglycerol (RRS)	8.2
Diricinoleoyloleoylglycerol (RRO)	5.6
Diricinoleoyllinoleoylglycerol (RRL)	1.2
Diricinoleoylpalmitoyl-glycerol (RRP)	0.9

Source: Salimon et al. [34].

Table 3.
Major triacylglycerols and their composition in castor seed oil.

9. Physical and chemical properties of seed oil

The physical and chemical properties of castor oil includes moisture content, density, refractive index, fire point, flash point, smoke point, cloud point, pour point, viscosity, color, pH, turbidity, lipid content, free fatty acids, acid value, saponification value, unsaponifiable matter, peroxide value, iodine value, cetane number, and calorific value, and their probable range is presented in **Table 4**. The difference in the value of these properties may be due to environmental factor which influences the growth and productivity of the seed. The moisture content of the crude oil lies between 0.2 and 0.31%, which indicates low moisture content that is the characteristics of good shelf life. The density ranges between 0.946 and 0.950 g/cm^3, which can be further reduced by esterification for application as biodiesel. The refractive index indicates the level saturation of the oil. The fire point, flash point, smoke point, cloud point, and pour point give evidence of good combustion quality as biofuel. The viscosity range (0.305–0.545 cps) indicates that the oil is light and highly unsaturated. The low levels of pH (5.8) notify the presence of modest amount of free fatty acid in the oil, which is a good indicator for utilization of oil in soap making. The free acids and acid value express the level of oxidative deterioration of the oil through enzymatic or chemical oxidation. However, the fatty acids can be transformed to edible oil through refining of crude oil and will also improve its quality for industrial use. The saponification value (182.9–327.4 mgKOH/g) expresses the relative length of fatty acid chain.

Properties	Average value range
Moisture (%)	0.2–0.31
Density (g/cm^3)	0.946–0.950
Refractive index at 25 C	1.47
Fire point (°C)	254.8–257.2
Flash point (°C)	222.9–227.1
Smoke point (°C)	214–216
Cloud point (°C)	3
Pour point (°C)	2
Viscosity (cps)	0.305–0.545
pH	5.8
Turbidity	4–6
Lipid content (%)	43.3–47.8
Free fatty acids	3.4–7.21
Acid value (mg/g)	4.9–14.42
Saponification value (mgKOH/g oil)	182.9–327.4
Unsaponifiable matter	3.4
Peroxide value (meq/kg)	10.2
Iodine value (wijj's value)	57.93–59.35
Cetane number	55.9
Calorific value (MJ/Kg)	36.25

Source: Salimon et al. [34]; Nangbes et al. [37]; Bale et al. [38]; Saifullah et al. [39].

Table 4.
Physicochemical properties of castor seed oil.

Abayeh et al. [40] reported that oil with high saponification value could be used as raw materials for soaps and cosmetics. Iodide value could be used to determine the total number of double bond present in the oil, which indicates the susceptibility of oil to oxidation. The peroxide value appraises the rancidity of the oil during storage process. Cetane number indicates the ignition quality, and calorific value represents the measures of available energy of fuel. All of these physical and chemical properties of castor oil established it as a good source of lubricant and biofuel and to be used for industrial purpose.

10. Toxic properties

The endosperm of castor seed contains a group of closely related toxic glycoproteins (ricin), ricinoleic acid, and the alkaloid ricinin. The seed cake of castor contains the toxic compound ricin, but castor oil does not contain ricin because it is insoluble in oil, and if remains it can be expelled in the refining process [3]. The toxic properties of castor seeds had been noticed since ancient times and its toxicity to human has recently been reported [41, 42]. Castor seed were used in classical Egyptian and Greek medicine and were delineated in Sanskrit medicine (Sushruta Ayurveda) from the sixth century BC [41]. More than 750 cases of accidental or deliberate intoxication have been reported in human [43]. The lethal dose in adults may be considered as four to eight seeds, but children are susceptible to small amount of seeds. An acceptable rate (0–0.7 mg/kg body weight) of daily castor oil consume for man has been established by the combined Food and Agriculture Organization (FAO) and World Health Organization (WHO) Expert Committee on Food Additives. Ricin creates health problem by damaging the ribosomes, which produce all of the protein needed by a cell and if the proteins cannot be produced cell may dies [4]. The symptom of ricin ingestion may be appeared up to 36 hours but generally start within 2–4 hours. This symptom includes a burning sensation in the mouth and throat, abdominal pain, diarrhea, fever, nausea, vomiting, incoordination, drowsiness, and hematuria. Severe dehydration, drop in blood pressure, and reduced urination have occurred within several days. If immediate treatment is not taken, vascular collapse and death may occur within 3–5 days; however, in most patients, full recovery can be possible [43].

The other toxic protein in castor seed is RCA (*Ricinus communis* agglutinin) which agglutinates red blood cells. When RCA is injected into the blood stream, it causes a person's blood to coagulate [4]. Toxicity also occurs in animals when they ingest broken seed or break the seed by chewing, but intact seed does not release toxin, it passes through the digestive tract [44]. Toxicity may vary to different animal at different dose. The toxin produced from castor seed is also used as natural insecticide and fungicide.

11. Uses of castor oil and its derivatives

11.1 Medicinal and pharmaceuticals

11.1.1 Antimicrobial activity

Ricinus communis shows antimicrobial activity against dermatophytic and pathogenic bacterial strains such as *Streptococcus progenies, Staphylococcus aureus*, as well as *Klebsiella pneumonia* and *Escherichia coli*. The result revealed that the petroleum ether and acetone extract inhibit microbial activity, whereas ethanolic extract has antibacterial activity only on higher concentration [45]. Different solvent extracts of root of *Ricinus communis* (200 mg/ml) occupy antimicrobial activity by utilizing well diffusion method against pathogenic microorganism such as *Escherichia coli*, *Staphylococcus aureus*, *Pseudomonas aeruginosa*, *Salmonella typhimurium*, *Proteus vulgaris*, *Bacillus subtilis*, *Candida albicans*, and *Aspergillus niger*. The hexane and methanol extracts possess highest antimicrobial activity, while aqueous extracts have no significant antimicrobial activity [46].

11.1.2 Antioxidant activity

It is concluded that seed extract of castor shows significant antioxidant activity by using lipid peroxidation method by ferric thiocyanate and free radical scavenging effect on 2,2-diphenyl-1-picrylhydrazyl radical (DPPH) and hydroxyl radicle produced from hydrogen peroxide. Those diseases which are caused by oxidative stress can be reduced by high antioxidant function of *Ricinus communis* seeds at low concentration. Methyl ricinoleate, ricinoleic acid, 1,2-octadecadienoic, and methyl ester are the chemical compound present in castor for which it shows antioxidant activity [47]. The leaf and stem extract of castor contains flavonoids, which show antioxidant activity [48, 49].

11.1.3 Antiulcer activity

The oil that is extracted from the seed of *Ricinus communis* has the potentiality to prevent ulcer at a dose of 500 mg/kg and 1000 mg/kg, but 1000 mg/kg is more effective against ulceration caused by pylorus ligation, aspirin, and ethanol in rats.

It is shown that the antiulcer activity of *Ricinus communis* is due to the cytoprotective action of drug or corroborant of gastric mucosa that ameliorate the mucosal protection [50].

11.1.4 Antidiabetic activity

Ethanolic extract of root of *Ricinus communis* shows significant effect in reducing the glucose level of fasting blood. An experiment was conducted on diabetic rats that showed that the glucose level was reduced from an initial level of 386 ± 41 mg/dl to 358 ± 33, 293 ± 28, 191 ± 25, 133 ± 29, 96 ± 20, and 79 ± 16 mg/dl on the 2nd, 5th, 7th, 10th, 15th, and 20th days, respectively. The fasting blood glucose became normal on the 20th days. Castor showed statistically similar result in alkaline phosphatase, serum bilirubin, creatinine, serum glutamate oxaloacetate transaminase, serum glutamate pyruvate transaminase, and total protein which was observed even after the administration of the extract at a dose 10 g/kg body weight. The extraction is also effective for total lipid profile and liver and kidney functions. Thus, *Ricinus communis* is considered as a powerful phytomedicine for diabetics [51].

11.1.5 Anti-inflammatory and free radical scavenging activity

The methanolic extract of castor root possesses anti-inflammatory and free radical scavenging activity. It was studied in Wistar albino rats in which oral administration of methanolic extract at a dose of 250 and 500 mg/kg body weight showed significant ($p < 0.001$) anti-inflammatory activity in carrageenin-induced hind paw edema model. The oral administration of the extract at the dose 500 mg/kg body weight also showed significant ($p < 0.001$) anti-inflammatory activity in cotton pellet granuloma model. The methanolic extract also showed free radical scavenging activity by suppressing lipid peroxidation initiated by carbon tetrachloride and ferrous sulfate in rat liver and kidney homogenates. The extract augments the free radical scavenging activity of stable radical 2,2-diphenyl-1-picrylhydrazyl radical (DPPH), nitric oxide, and hydroxyl radical in in vitro assay methods [52].

11.1.6 Central analgesic activity

The crude extract of root bark of castor possesses analgesic activity in tail flick response model to radiant heat at a dose of 250 mg/kg [53]. The ethanolic extract of fruit pericarp of castor contains typical CNS stimulant and neuroleptic effects. The stimulant effects such as exophthalmus, hyperreactivity (evidence by tremors or by the pinna and grip strength reaction), memory improvement, and clonic seizures seem to be due to the presence of alkaloid ricinine, which is considered as a main toxic compound in the extract. Animals that died after being treated with extract showed similar signs; they all died by clonic seizures, which were followed by apparent breathing arrest. On the other hand, compounds other than ricinine may cause neuroleptic-like effects of extract because ricinine is not responsible for reducing locomotor function or catalepsy in mice [54].

11.1.7 Anticancer activity

In spite of being a poisonous compound, ricin possesses the potentiality to prevent tumor and has been used in cancer research and chemotherapy during recent years. One of the most important uses of ricin is in the manufacturing of immunotoxins where the protein ricin is jointed to monoclonal antibodies. In vitro technique was used to produce these antibodies which have a protein receptor site that identify the specific target cells in the tumor. The protein antibody joined together and resulting compound is known as immunotoxin. For the treatment of a cancer patient, the deadly toxin can be carried at the site of tumor by arming these antibodies with ricin. Thus, castor has the most significant use in the treatment of tumor or cancer cell due to the presence of ricin. It kills only the target tumor cell without damaging other in the patient [55].

11.1.8 Antifertility activity

The methanolic extract of seeds of castor plant has a positive preliminarily phytochemical test for steroid as well as alkaloids. The pituitary gland releases gonadotropins by both positive and negative feedback mechanism due to sex hormones, and also the pituitary gland inhibits the release of luteinizing hormone and follicle-stimulating hormone. This is due to estrogen and progesterone fall a cumulative effect in luteal phase of menstrual cycle. Thus, it helps the inhibition of maturation follicle in the ovary and prevents ovulation. As sex hormone is a steroidal compound, methanol extract of castor seed contains steroids; it produces antifertility effect [56, 57].

11.1.9 Lipolytic activity

The compound ricin show lipolytic function by using different substrates such as (a) one analogue of triacylglycerol, BAL-TC_4 (b), different chromogenic substrates such as *p*-NP esters of aliphatic short to medium chain acids, and (c) monomolecular films of a pure natural diacylglycerol, DC_{10}, in emulsion and in a membrane-like model. An experiment was conducted that revealed that ricin of *Ricinus communis* acts as lipase and possesses the ability to hydrolyze different lipid classes as well as phospholipid (important constituent of cellular membrane). The lipolytic activities are maximum at pH 7.0 in the presence of 0.2 M galactose. The action of ricin on membrane phospholipid occurred by phospholipase A1 activity which may be regarded as minor activity of lipase [58].

11.1.10 Hepatoprotective activity

The ethanolic leaf extract of castor at 250/500 mg/kg body weight possesses hepatoprotective activities. This is because of their inhibitory activities of an increase in the function of serum transaminase, level of liver lipid per oxidation, protein, and glycogen, and the activities of acid and alkaline phosphatase in the liver initiated by carbon tetrachloride (CCl_4) are increased. The extract also treated the depletion of glutathione level and adenosine triphosphatase functions in the CCl_4-induced rat liver. Flavonoids are present in the ethanolic extract which possess membrane-stabilizing and antiperoxidative effects. *Ricinus communis* also increases regenerative and reparative ability of liver because of the presence of flavonoids and tannins. Duo to the presence of N-demethyl ricinine in the leaves of *Ricinus communis*, it showed anticholestatic and hepatoprotective activity against paracetamol-induced hepatic damage. The whole leaves of *Ricinus communis* have the potentiality to defend against liver necrosis and fatty changes generated by CCl_4, while the glycoside and cold aqueous extract give protection against liver necrosis and fatty changes, respectively [59–62].

11.1.11 Wound healing activity

Castor oil is effective for injury healing. It produces antioxidant activity and inhibits lipid peroxidation. The compound, which is responsible for inhibiting lipid per oxidation, is supposed to increase viability of collagen fibrils by increasing the strength of collagen fibers, increasing the circulation, and preventing cell damage by promoting DNA synthesis. Tannins, flavonoids, triterpenoids, and sesquiterpenes have astringent and antimicrobial properties, which repair wound portion and increase the epithelialization. An observation showed that castor oil healed the wound area by reducing the scar area and epithelialization time in excision wound model. A comparison study showed that 10% w/w concentration has better wound healing property than 5% w/w concentration [63].

11.1.12 Anti-asthmatic activity

There are some compounds present in the root extract of castor plant which is very important for the treatment of asthma, because it has anti-allergic and cell-stabilizing ability. Saponins present in the extract possess mast cell-stabilizing effect, and flavonoids have smooth muscle relaxant and bronchodilator function. The apigenin- and luteolin-like flavonoids normally restrict basophil histamine release and neutrophils beta glucuronidase release and show anti-allergic activity in vivo. Another study showed that ethanolic extract of *Ricinus communis* reduces milk-induced leukocytosis and eosinophilia and possesses anti-asthmatic activity as flavonoids and saponins are present [64].

11.1.13 In vitro immunomodulatory activity

The immunomodulatory compound generally increases the immune response of the human body against different pathogens by activating the nonspecific immune system. The phagocytosis is the engulfment of microorganisms by leucocytes which is one of the main protective mechanisms of the organism. The final step of phagocytosis is the intracellular killing of microorganism by the neutrophils. The leaves of *Ricinus communis* contain tannins, which increase the phagocytic function of human neutrophils and thus produce immunomodulatory effect [65].

11.1.14 Bone regeneration activity

Experiments were conducted to see the biocompatibility and potentiality of *Ricinus communis* polyurethane (RCP) to regenerate the bone. Result revealed that *Ricinus communis* polyurethane is mixed with calcium carbonate and phosphate which promote matrix mineralization and are biocompatible materials [66]. The biological properties of RCP are improved by incorporating alkaline phosphate to it with subsequent incubation in synthetic body fluid [67]. The benefit observed in RCP as compared to demineralized bone is that the former has slow reabsorption process [68].

12. Lubricants, hydraulic, and brake fluid

Castor oil has been used to develop low pour point lubricant base stocks by synthesizing acyloxy castor polyol esters [69]. Due to having low pour point property, it provides full lubrication to the equipment in cold environment [70]. An interesting study by Singh revealed that castor oil-based lubricant has the luscious potentiality to be used as smoke pollution reducer. In his research he used a biodegradable two-stroke (2 T) oil, which is a popular variety of lubricating oil and was used on two-stroke engines in scooters and motorcycles. The lubricant comprises tolyl monoesters and performance additives but no miscibility solvents. The result revealed that it decreased smoke by 50–70% at a 1% oil/fuel ratio [71]. Castor oil also can be used as car engine lubricant. A modified version of castor oil lubricant comprising 100 parts of castor oil and 20–110 parts of a chemically and thermally stable, low viscosity blending fluid, soluble in castor oil showed its potential to be used as a lubricant for refrigeration system [72]. In spite of having its use as DOT 2 rating brake fluid, castor oil lubricant is considered as obsolete types of brake fluid and is not used in the modern vehicles [73].

13. Food

Food grade castor oil is used in the food industry. It can be used as food additives such as flavor and food color and as a mold inhibitor and in packaging. In the food-stuff industry, polyoxyethylated castor oil is also used [74]. The white, large seed of castor are an important source of food condiment called "Ogiri" in the southeastern part of Nigeria [75].

14. Polymer materials

For synthesizing the renewable monomers and polymers in the castor oil and its derivatives are used [76]. To produce the vulcanized and urethane derivatives, castor oil was polymerized with sulfur and diisocyanates, respectively [77]. In other study, by sequential mode of synthesis, full-interpenetrating polymer networks (IPNs) were prepared using epoxy and castor oil-based polyurethane (PU) (Raymond and Bui [78]). Similar to the aforementioned study, a series of two components IPN of the modified castor oil-based PU and polystyrene were prepared by sequential method [79]. IPN is also known as polymer alloy and is considered as one of the fastest growing research areas in the field of polymer blends in the last two decades [79]. As a root-end filling material, castor oil polymer has been shown to possess sealing ability. Root-end filling material is the root-end preparation filled with experimental materials, and it provides an apical seal to prevent the bacterial movements and its diffusion from root canal system to peripheral tissues [80]. One of the most common applications using castor oil is biodegradable polyesters [81]. The first synthetic condensation polymers are polyesters which are environmentally safe and friendly. This is also useful in biomedical field as well as elastomers and packaging materials [82, 83]. Castor can be combined with other monomers with a view to produce an array of copolymers. Again these copolymers provide materials with different properties which find use in products ranging from solid implants to in situ injectable hydrophobic gel [81].

15. Preparation of soap, waxes, and greases

Castor oil is used to produce soaps and waxes [30]. In a study by Dwivedi and Sapre [84], they utilized castor oil in total vegetable oil greases. Total vegetable oil greases are those in which both lubricant and gallant are formed from vegetable oil. In their study, they utilized a simultaneous reaction scheme to produce sodium and lithium greases from castor oil.

16. Coating

Castor oil can be used in producing coatings and paints. For useful paintings and furniture oil application, castor oil is dehydrated by monoconjugated oil-maleic anhydrite adducts [85]. Castor oil is utilized as coating application by converting the hydroxyl functionalities of castor oil to β-ketoesters using *t*-butyl acetoacetate [86]. Advanced surface coating materials were synthesized from castor oil-based hyperbranched polyurethanes (HBPUs) which is a highly branched macromolecule [87]. Most recently, Allauddin et al. [88] synthesized a high-performance hybrid coating by using a methodology that consists of introducing hydrolyzable-Si-OCH_3 groups onto castor oil that have been used for the development of PU/urea-silica hybrid coating.

17. Fertilizer

There are two by-product produced from the castor seed, i.e., husks and meal. Lima et al. [89] reported that the blend of castor meal and castor husks can be used as fertilizer, which is effective for substantial plant growth when it is applied up to a dose of 4.5% (in volume) of meal. But when the dose exceeds 4.5%, the plant growth is retarded and even the plant may die [89].

18. Other uses

Besides the above uses, the *Ricinus communis* is used for different purposes. The oil is used in coating fabrics and other protective covering, in the production of typewriter and printing inks. Castor oil is also used in textile dyeing. The hydrogenated oil is useful for the production of polishes, carbon paper, candles, crayons, etc. The cellulose from stem is used to prepare cardboard, paper, etc. Polyoxyethylene hydrogenated castor oil is also useful for the manufacture of vitamin A and vitamin C, eye drop, and oral nitroglycerine sprays [90].

19. Biodiesel production from castor oil

19.1 Castor oil extraction

The extraction of castor oil from castor seed can be done by either mechanical pressing or solvent extraction or a combination of both. After harvesting, the seeds are dried to split open the seed hull so that kernel can be collected easily. Extraction process starts with the dehulling of seeds, and this can be done either manually by hands or mechanically with the help of a castor seed dehuller. After dehulling the seed, foreign materials such as sticks, stems, leaves, sand or dirt are removed by using a series of revolving screens or reels. After cleaning, the kernels are heated in a steam jacketed press to eliminate moisture, and then these cooked kernels are dried; this hardening will help in extraction.

19.1.1 Mechanical extraction

A hydraulic press or oil expeller is used to remove oil from castor kernels. This mechanical extraction is done at low temperature which recovers only about 45% oil from the castor seeds. Higher temperature can increase the extraction efficiency up to 80% of the available oil which can be done by using high-temperature hydraulic press. The extraction temperature can be maintained by circulating of cold water through the pressing machine that is responsible for cold pressing of kernels. Cold-pressed castor oil contains low acid and iodine content and is lighter in color than the castor oil which is solvent extracted. After extraction, the oil is collected and filtered, and the filtered materials are mixed with fresh kernels for repeat extraction. The extraction process is repeated for several times by bulking of filtered material with new material and oil is collected. The by-product is finally removed from the press as seed cake. This seed cake contains about 10% of castor oil [1]. The remaining oil in seed cake can be obtained by crushing the seed cake and subjected to solvent extraction.

19.1.2 Solvent extraction

The solvent extraction of castor oil can be done by using Soxhlet extractor. About 300 ml of solvent such as hexane, heptane, or petroleum ether is poured in a round-bottom flask, and 10 g of crushed castor kernel packed with oil tissue or filter paper is placed in a thimble and inserted into the center of the extractor. The extractor then is fixed on the round-bottom flask, and a condenser is placed on the top of the extractor. Then the fitted apparatus is placed in a heating mantle and heated (50–60°C) to boil the solvent. When the solvent starts to boil, the vapor rises through the vertical tube into the condenser at the top. The vapor condensed and dripped into the thimble at the center. The extract seeps through the pores of the thimble and fills the siphon tube where it flows back down into the round-bottom flask [91]. The extraction process is continued for 8 hours, and after that, the extract with solvent in the round-bottom flask is subjected to rotator evaporator to recover the solvent from the extracted oil. The weight of extracted oil should be recovered for further determination.

19.2 Transesterification

The transesterification process is the reaction of a triglyceride with an alcohol to produce ester and glycerol. A triglyceride has a glycerine molecule as its base with three long chain fatty acids annexed. The characteristics of the fat are determined by the nature of the fatty acids subsumed to the glycerine which affects the characteristics of the biodiesel. In the production of biodiesel, vegetable oil in the form of triglycerides reacts with small chain alcohol (methanol, ethanol, propanol, etc.) in the presence of homogeneous catalyst such as base (KOH, NaOH) or acid (HCl, H_2SO_4, H_3PO_4) or heterogeneous catalyst as zeolites or biocatalyst as enzymes. The process is also called alcoholysis. When methanol is used, it is called methanolysis, and esters that are produced in methanolysis are called fatty acid methyl esters (FAMEs), and in case of ethanol, the process is termed as ethanolysis, and the esters produced in this process are called fatty acid ethyl esters (FAEEs) [92]. The transesterification is a reversible reaction, so alcohol must be added in excess to ensure the reaction in the right direction (**Figure 3**).

19.2.1 Procedure

For transesterification about 25 ml of oil was kept in three-necked round-bottom flask and heated to 65°C. Then, the required quantity of methanol and catalyst (KOH) is added with stirring system. The experiment was continued for 3 hours and then the sample was monitored by running TLC to ensure the completion of reaction. After cooling, two layers were differentiated by separatory funnel, the

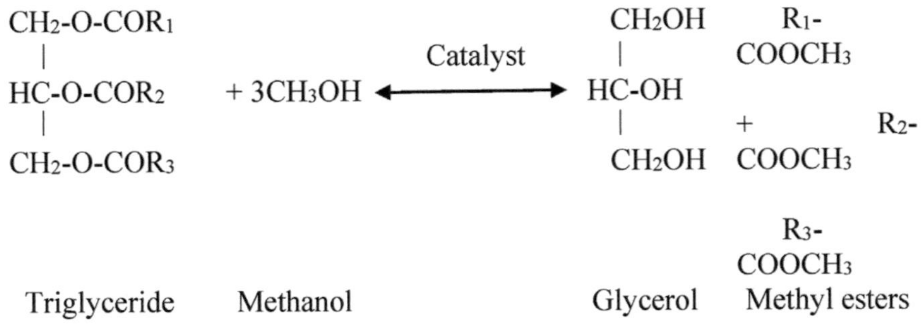

Figure 3.
Transesterification reaction.

upper layer is methyl ester (biodiesel), and the lower layer is glycerol. Produced methyl ester could be purified by successive rinse with 2.5% (w/w) H_2SO_4 and distilled water. NaCl was used to avoid emulsion during washing process. The washed methyl ester should be treated with anhydrous sodium sulfate to eliminate excess water. It was then filtered and dried by heating at low temperature (60°C) for 30 minutes [93].

19.3 Filtration or purification

After extraction of oil by using oil expeller, there still remain impurities in the extracted oil which can be removed through filtration process. Large- and small-sized particulates, any dissolved gases, acids, or even water can be removed by using filter press. Crude oil of castor seed is pale yellow or straw in color, but it can be made colorless or near colorless by refining and bleaching. The crude of castor seed also has a discrete odor which can also be deodorized during refining process [1].

19.4 Refining

Following filtration of crude oil of castor, it is subjected to refining process to eliminate impurities such as colloidal matter, phospholipids, excess free fatty acids, and coloring agents. Removal of these impurities prevents deterioration during long-term storage. The refining process includes several steps such as degumming, neutralization, bleaching, and deodorization, and sometimes winterization should be performed for efficient oil refining [1].

19.4.1 Degumming

Degumming is performed to reduce the phospholipids and metal content of the crude oil of castor. The forms of phospholipids found in crude castor oil are lecithin, cephalin, and phosphatidic acids [94], and these phosphatides can be classified as hydratable and nonhydratable [95]. For efficient removal of these phosphatides, a suitable degumming procedure such as water degumming, acid degumming, and enzymatic degumming has to be implemented. Generally crude vegetable oil contains about 10% of nonhydratable phosphatides [95] which may vary depending on several factors such as type of seed, quality of seed, and condition applied during milling operation. Water degumming process can be followed to remove hydratable phosphatides, and nonhydratable phosphatides can only be eliminated by applying acid or enzymatic degumming procedures [95].

19.4.2 Neutralization

Neutralization is the process of removing excess free fatty acids (FFAs) from the degummed oil. The FFA content is high in old seeds, which are stored for more than 1 year with high moisture content [96]. The degummed castor oil is refined by chemical refining or alkali neutralization which abates the content of FFAs, oxidation products of FFAs, residual proteins, phosphatides, carbohydrates, traces of metals, and a part of pigments. The alkali neutralization is done by treating degummed castor oil with an alkali solution (2% caustic soda) at temperature between 85 and 95°C with continuous stirring for about 45–60 minutes [97]. At this stage the alkali reacts with FFAs and converts them into soap which has a higher specific gravity than the neutral oil and tends to settle at the bottom. The oil can be differentiated either by gravity separation or by using commercial centrifuges. The separated oil is then washed with hot water to remove soap, alkali solution, and other impurities [98]. For batch neutralization of castor oil, it needs four to six times hot water wash so as to reduce the level of soap below 100 ppm [97]. The oil, thus obtained, is dried in vacuum dryer and transferred to the bleaching process.

19.4.3 Bleaching

After degumming and neutralization, the castor oil that appeared is clear and liquid, but it may still contain colored bodies, natural pigments, and antioxidants (tocopherols and tocotrienols). However, bleaching, an adsorption process, is used to remove such colored pigments and phospholipids. Bleaching of castor oil can be done under vacuum at about 100°C, and continuously stirring the oil with appropriate amount activated earths and carbon [91]. The activated earths are clay ores that consist of minerals such as bentonite and montmorillonite. About 2% bleaching earth and carbon are required in the bleaching process to produce desirable light-colored oil. In this process, colored particles, soap, and phosphatides are adsorbed by the activated earth and carbon. A commercial filter is used to remove the activated earth and carbon. The spent earth and carbon thus obtained contains about 20–25% oil content [99]. This retained oil in earth can be recovered by boiling the spent earth in water or by solvent extraction method. The oil that is recovered from the spent earth is highly colored with high FFA and high peroxide content usually more than 10 mg KOH/g and 20 meq/kg, respectively [100].

19.4.4 Deodorization

Deodorization is vacuum distillation processes that carry away relatively volatile components that produce undesirable flavor, color, and odors in fat and oils. To produce pharmaceutical grade castor oil, deodorization is necessary, but in other cases, this process is not essential as it is a nonedible vegetable oil [101, 102]. Deodorization is generally under high vacuum and temperature above 250°C to expel undesirable odor caused by ketones, aldehydes, sterols, triterpene alcohols, and short chain fatty acids [98]. Pharmaceutical grade castor oil is deodorized under low temperatures (150–170°C) and high vacuum for 8–10 hours to hydrolysis of hydroxyl group ricinoleic acid [103].

19.4.5 Winterization

Most vegetable oils contain high concentration of waxes, fatty acids, and lipids which is subjected to winterization before final use. Winterization is the process where waxes are crystalized and eliminated by a filtering process to avoid clouding of liquid fraction at cooler temperatures [1].

20. Challenges

20.1 Development of high-yielding varieties

For the development of high-yielding varieties of castor understanding, the genetics of economically important traits is most important. Different morphological and qualitative traits are controlled by one or few genes and their additive, dominant, and epistatic effects, which make it more difficult to develop high-yielding varieties. Stem color of castor is controlled by epistatic interaction of two genes "M" and "G" [104] and tall plant shows dominance over dwarf plant due to a monogenic factor. Particularly the inheritance of sex expression is important in the development of hybrids. There are three types of pistillate line, i.e., N, S, and NES, which could be used for hybrid production. In the N type, the occurrence of only female flowers is controlled by a recessive gene (ff); in the S type, the production of only female flower is controlled by a polygenic complex with dominant and epistatic effects; and in the NES type, the induction of female is also controlled by a recessive gene (ff), but sexual reversion occurs when the air temperature is more than 31°C [105–107]. The seed yield and seed oil content are usually inherited by quantitative manner. Some important characters such as the number of nodes before flowering, number of racemes per plant, and seed oil content are controlled by additive genetic effect [108, 109]. Other traits such as length of primary raceme, number of capsules per racemes, and seed weight are also additively inherited [110–112]. Early maturity is an another important character for castor cultivation in tropical areas or regions of short growing seasons where multiple crops are cultivated, but it shows negative correlation with high seed yield which is the main hindrance in the development of early maturing variety [113]. Genetic transformation of castor also remains challenging as it is averse to proficient regeneration of durable and transformed plant. The callus culture of castor for regeneration of plant has been problematic due to the lack of proper protocol which restricted the development of transgenic cultivars [114]. The most important global challenge in castor breeding is the development of cultivars that facilitate mechanical harvest. The success of perennial and indeterminate type castor is limited than annual and determinate type. The selection dwarf and non-branching type castor plant is hardly possible due to high genotype versus environment interaction.

20.2 Disease and pests

The most important challenges in castor cultivation are management of disease and pest incidence. Several disease occurrences were noticed in castor; among these gray mold (*Botryotinia ricini*), vascular wilt (*Fusarium oxysporum* f. sp. *ricini*), and charcoal rot (*Macrophomina phaseolina*) are the major diseases. Some other disease causes epidemic condition depending on the genotype and environmental conditions such as the leaf spot caused by the fungus *Alternaria ricini* and *Cercospora ricinella* and the bacteria *Xanthomonas axonopodis* pv. *ricini*. Among these *Alternaria ricini* is the most important because it is a seed-borne disease and causes seedling blight and pod rot with the loss of seed yield up to 70% [3]. Several plant parasitic nematodes are noticed on castor, but they do not cause severe damage [115]. Among these reniform nematodes, *Rotylenchulus reniformis* is the most important because it predisposes castor to the infection of *Fusarium oxysporum* [116]. Gray mold is considered as the most serious disease worldwide, but a few studies have been conducted recently on this disease [117]. Resistant varieties cannot be developed through breeding programs, but a few genotypes moderately tolerant to this disease have been identified [5]. Further studies have been needed for the management *Botryotinia ricini*. The occurrence of vascular wilt can be managed through varietal resistance, seed treatment, and crop rotation. Charcoal rot or *Macrophomina* root rot can be managed through cultivar resistance, but crop rotation and organic matter rectification can abate the severity of this disease [118].

The major insect pests that cause significant damage are castor semilooper (*Achaea janata*), castor shoot borer (*Conogethes punctiferalis*), capsule borer (*Dichocrocis punctiferalis*), tobacco caterpillar (*Spodoptera litura*), red hairy caterpillar (*Amsacta* spp.), and leaf miner (*Liriomyza trifolii*) [16, 119]. In Brazil, the major insect pests of castor are stink bug (*Nezara viridula*); leafhopper (*Empoasca* spp.); defoliator including armyworm (*Spodoptera frugiperda*), semilooper (*Achaea janata*), and black cutworm (*Agrotis ipsilon*); and the mites *Tetranychus urticae* and *Tetranychus ludeni* [120, 121]. Cotton lace bug (*Corythucha gossypii*) was also noticed as a pest of castor in Colombia. The integrated pest management program with pesticides and crop rotation, insect traps, neem extract can be used to manage the insect pests of castor [119].

21. Detoxification of castor product

After the extraction of oil from castor seed, it produces castor meal as a by-product, which contains a toxic compound ricin. This ricin content is about 1 to 5% of the weight of the castor meal remaining after oil extraction [122, 123]. Small quantities of castor meal can be easily detoxified, but no commercial or industrial level detoxification process has been successfully implemented yet. In early 1934, it was demonstrated that by boiling for 2 hours, castor meal could be detoxified. Several other methods for castor meal detoxification have been investigated later which includes short but repeated boiling, autoclaving, steam heating, fermentation, ionizing radiation, and mixing castor meal with tannin-rich meal of Sal seed (*Shorea robusta*) and the addition of sodium hypochlorite, alkali, or acid substances [3]. Both ricin and allergens was detoxified simultaneously by adding calcium hydroxide followed by extrusion [124]. At present, the addition of lime is the simplest and effective method of ricin detoxification. Probably the high pH is responsible for the denaturation of ricin [125–127]. The economics and access to commercial castor production can be improved through the development of industrial process of castor detoxification, but the impediment is high-energy costs for processing the meal, decreased in feed quality of processed meal, and the absence of proper methods to promptly and cheaply quantify the residual ricin in the meal [3].

22. Present status

According to FAOSTAT, during 2014, the average world production of castor oilseed was 1.95 million tons that was harvested from an area of 1.44 million hectare, of which 92.2% was concentrated in Asia, mostly in India. India ranked first in the production of castor oil seed that was about 1.73 million tons followed by Mozambique (0.069 million ton), China (0.04 million ton), Brazil (0.038 million ton), and Myanmar (0.011 million ton). In Bangladesh, it is only 266 tons, which is too much lower compared to India, Mozambique, and China. India is the highest exporter of castor oil accounting for more than 90% of the castor oil exports, while the United States, European countries, and China are the major importer, accounting for more than 84% of the imported castor oil [128]. Harvested area, production, and yield of the top 10 castor oil seed producers during 2014 are presented in **Table 5**.

23. Future prospects

The global consumption of castor is increased, but the current production of castor is not increasing at sufficient rate. Future research strategies play an important role in the world production of castor. International collaboration between scientific communities is needed for the development of solution to the main constraints to castor production, processing, and marketing. Although some locally adapted variety and hybrids are developed, an integrated plant improvement strategy needs to be developed for further progress. A closer interaction between plant breeders, molecular biologists, plant pathologists, plant physiologists, and entomologists is needed for speeding up the research activities. Both the quality and quantity of castor oil can be improved by using biotechnological innovations and genetic engineering. The castor genome draft should be used as map for introducing molecular markers in castor breeding. Improved coordination of germplasm bank helps in the standardization of evaluation method and increase the exchange of accession in breeding programs. The development of non-shattering, dwarf, and high-yielding cultivar with additional improvement in machinery and agronomic practices will allow the prompt transition of castor to mechanized production [129]. Breeding of castor for resistance or tolerance to disease and insect pest is also important for the production of good-quality castor oil seed. Another major concern is the development of castor cultivar with low ricin, low ricinin, low allergen, and low RCA content [41, 42, 130]. Accurate detection and detoxification of castor toxin in feed and biological samples remain a challenge to the commercial use of castor meal in animal rations. The use of castor oil for biodiesel production is problematic due to its high viscosity and high cost of production and refining. However, castor has a tremendous potentiality as a source of bioenergy and industrial feedstock with high oil content, unique fatty acid composition (ricinoleic acid), and a wide range of adaptation under drought and saline condition.

Country	Harvested area (million hectare)	Production (million ton)	Yield (ton/ha)
India	1.04	1.733	1.666
Mozambique	0.184	0.069	0.375
China	0.046	0.040	0.870
Brazil	0.063	0.038	0.591
Myanmar	0.014	0.011	0.782
Ethiopia	0.005	0.011	2.000
Paraguay	0.008	0.009	1.125
Vietnam	0.008	0.007	0.875
South Africa	0.010	0.006	0.607
Angola	0.016	0.004	0.253

Source: FAO [2].

Table 5.
Harvested area, production, and yield of the top 10 castor oil seed producers during 2014.

24. Conclusion

Castor is an underutilized nonedible oil crop species that has a variety of application, but it is promising for its high oil content particularly as a potential source of renewable energy. It is also used in the production of pharmaceuticals, lubricants, hydraulic and brake fluid, polymer materials, coating, and fertilizer. It also contains toxic compounds that are ricin, ricinin, and RCA. The development of high-yielding varieties, detoxification castor meal, and control of insect pests are the major challenges.

Acknowledgements

The authors acknowledge the support of Bangabandhu Sheikh Mujibur Rahman of the Agricultural University, Gazipur, Bangladesh, for providing all research inputs and bearing the cost of the project. The authors would like to thank BSMRAU authority for their support.

Conflict of interest

There is no conflict of interest regarding the publication of the chapter.

Author details

Swapan Chakrabarty[1], Abul Kalam Mohammad Aminul Islam[1*], Zahira Yaakob[2] and Abul Kalam Mohammad Mominul Islam[3]

1 Faculty of Agriculture, Department of Genetics and Plant Breeding, Bangabandhu Sheikh Mujibur Rahman Agricultural University, Gazipur, Bangladesh

2 Faculty of Engineering and Built Environment, Department of Chemical and Process Engineering, Universiti Kebangsaan Malaysia (UKM), Bangi, Selangor, Malaysia

3 Faculty of Agriculture, Department of Agronomy, Bangladesh Agricultural University, Mymensingh, Bangladesh

*Address all correspondence to: aminuljkkp@yahoo.com

References

[1] Patel VR, Dumancas GG, Viswanath LCK, Maples R, Subong BJJ. Castor oil: Properties, uses, and optimization of processing parameters in commercial production. Lipid Insights. 2016;**9**:1

[2] FAO. FAOSTAT Data base. Rome, Italy: Food and Agriculture Organization; 2014

[3] Severino LS, Auld DL, Baldanzi M, Cândido MJ, Chen G, Crosby W, et al. A review on the challenges for increased production of castor. Agronomy Journal. 2012;**104**(4):853-880

[4] Helmenstine AM. Ricin and RCA. ThoughtCo; 2017. Available from: https://www.thoughtco.com/overview-of-castor-bean-toxins-602403 [Accessed: 01 July 2017]

[5] Anjani K. Castor genetic resources: A primary gene pool for exploitation. Industrial Crops and Products. 2012;**35**(1):1-14

[6] Govaerts R, Casas FF, Barker C, Carter S, Davies S, Esser HJ, et al. World Checklist of Euphorbiaceae. Kew: The Royal Botanic Gardens; 2014

[7] CABI. *Ricinus communis* (Castor Bean). 2016. Available from: http://www.cabi.org/isc/datasheet/47618 [Accessed: 21 June 2017]

[8] Maroyi A. *Ricinus communis*. 2007. Available from: http://uses.plantnet-project.org/en/Ricinus_communis_(PROTA) [Accessed: 15 June 2017]

[9] Rizzardo RA, Milfont MO, Silva E, Freitas BM. Apis mellifera pollination improves agronomic productivity of anemophilous castor bean (*Ricinus communis*). Anais da Academia Brasileira de Ciências. 2012;**84**(4):1137-1145

[10] Zimmerman LH, Smith JD. Production of F1 seed in castorbeans by use of sex genes sensitive to environment. Crop Science. 1966;**6**(5):406-409

[11] Shifriss O. Conventional and unconventional systems controlling sex variations in *Ricinus*. Journal of Genetics. 1960;**57**(2):361-388

[12] Brigham RD. Castor: Return of an old crop. In: Janick J, Simon JE, editors. New Crops. New York: Wiley; 1993. pp. 380-383

[13] Moreira J, Lima EF, Farias FJC, de Azevedo DMP. Melhoramento da mamoneira (*Ricinus communis* L.). Embrapa Algodão-Documentos (INFOTECA-E); 1996. Available from: https://www.infoteca.cnptia.embrapa.br/bitstream/doc/269457/1/MELHORAMENTODAMAMONEIRA.pdf

[14] Gangaiah B. Castor. 2008. Available from: http://nsdl.niscair.res.in/jspui/handle/123456789/530 [Accessed: 20 June 2017]

[15] Tomar Rukam S, Parakhia MV, Kavani RH, Dobariya KL, Thakkar JR, Rathod VM, et al. Characterization of castor (*Ricinus communis* L.) genotypes using different markers. Research Journal of Biotechnology. 2014;**9**(2):6-13

[16] Anjani K. Extra-early maturing germplasm for utilization in castor improvement. Industrial Crops and Products. 2010;**31**(1):139-144

[17] Moshkin VA, Dvoryadkina AG. Cytology and genetics of qualitative characters. In: Moshkin VA, editor. Castor. New Delhi, Amerind; 1986. pp. 93-102

[18] Uguru MI, Abuka LN. Hybrid vigour and gene action for two quantitative traits of castor plant

(*Ricinus communis* L.). Ghana Journal of Agricultural Science. 1998;**31**(1):81-86

[19] Bahia HF, Silva SA, Fernandez LG, da Silva Ledo CA, Moreira RFC. Divergência genética entre cinco cultivares de mamoneira. Pesquisa Agropecuária Brasileira. 2008;**43**(3):357-362

[20] Auld DL, Zanotto MD, McKeon T, Morris JB. Castor. In: Vollmann J, Rajcan I, editors. Oil Crops—Handbook of Plant Breeding. New York, USA: Springer; 2009. pp. 316-332

[21] Bertozzo F, Lara ACDC, Zanotto MD. Genetic improvement of castor bean in order to increase female flowers. Bragantia. 2011;**70**(2):271-277

[22] Fernández-Martínez JM, Velasco L. Castor. In: Gupta SK, editor. Technological Innovations in Major World Oil Crops. Vol. 1. Breeding. New York: Springer; 2012. pp. 237-265

[23] Filho AS. Melhoramento da mamona. In: Borem A, editor. Melhoramento de Especies Cultivadas (In Portuguese). Vicosa: Universidade Federal de Vicosa/Springer; 2005. pp. 385-407

[24] Amaral JGC. Genetic variability for agronomic characteristics between self pollinated lines of castor (Ricinus communis L.) cv. AL Guarany [PhD dissertation (Agronomy)]. Botucatu: College of Agronomic Sciences, São Paulo State University; 2003

[25] Kulkarni LG, Ramanamurthy GV. Castor (*Ricinus communis* Linn.). New Delhi: ICAR; 1977

[26] Zanotto MD, Amaral JGC, Poletine JP. Recurrent selection with use of self pollinated lines for reduction of the height of plants of castor (*Ricinus communis* L.) in a common Guarani population. In: Annals of the I Brazilian Congress of Mamona. 2004. pp. 1-5

[27] Oliveira IJ, Zanotto MD. Efficiency of recurrent selection for reduction of the stature of plants in castor (*Ricinus communis* L.). Ciencia e Agrotecnologia. 2008;**32**:1107-1112

[28] Yadava DK, Vasudev S, Singh N, Mohapatra T, Prabhu KV. Breeding major oil crops: Present status and future research needs. In: Technological Innovations in Major World Oil Crops. Vol. 1. New York: Springer; 2012. pp. 17-51

[29] Islam AKMA, Yaakob Z, Anuar N, Primandari SRP, Osman M. Physiochemical properties of *Jatropha curcas* seed oil from different origin and candidate plus plants (CPPs). Journal of American Oil Chemists' Society (JAOCS). 2011;**89**(2):293-300. DOI: 10.1007/s11746-011-1908-7

[30] Karmee SK, Chadha A. Preparation of biodiesel from crude oil of Pongamia pinnata. Bioresource Technology. 2005;**96**(13):1425-1429

[31] Shrirame HY, Panwar NL, Bamniya BR. Bio diesel from castor oil—A green energy option. Low Carbon Economy. 2011;**2**(1):1-6

[32] Kostik V, Memeti S, Bauer B. Fatty acid composition of edible oils and fats. Journal of Hygienic Engineering and Design. 2013;**4**:112-116

[33] Islam AKMA, Yaakob Z, Ghani JA, Anuar N. *Jatropha curcas* L.: A Future Energy Crop with Enormous Potential. Biomass and Bioenergy: Processing, Properties and Applications. 2014. pp. 31-61. DOI: 10.4236/lce.2011.21001. Published Online March 2011. Available from: http://www.SciRP.org/journal/lce

[34] Salimon J, Noor DAM, Nazrizawati AT, Firdaus MM, Noraishah A. Fatty acid composition and physicochemical properties of Malaysian castor bean *Ricinus communis* L. seed oil. Sains Malaysiana. 2010;**39**(5):761-764

[35] Conceição MM, Candeia RA, Silva FC, Bezerra AF, Fernandes VJ, Souza AG. Thermoanalytical characterization of castor oil biodiesel. Renewable and Sustainable Energy Reviews. 2007;**11**(5):964-975

[36] Gupta SS, Hilditch TP, Riley JP. The fatty acids and glycerides of castor oil. Journal of the Science of Food and Agriculture. 1951;**2**(6):245-251

[37] Nangbes JG, Nvau JB, Buba WM, Zukdimma AN. Extraction and characterization of castor (*Ricinus communis*) seed oil. International Journal of Engineering Science. 2013;**2**:105-109

[38] Bale AT, Adebayo RT, Ogundele DT, Bodunde VT. Fatty acid composition and physicochemical properties of castor (*Ricinus communis* L.) seed obtained from Malete, Moro local government area, Kwara State, Nigeria. Chemistry and Materials Research. 2013;**3**(12):11-13

[39] Saifullah AZA, Karim MA, Karim MR. Advancement of biodiesel in Bangladesh. IOSR Journal of Engineering. 2016;**6**(6):59-64

[40] Abayeh OJ, Aina EA, Okounghae CO. Oil content and oil quality characteristics of some Nigerian oil seeds. Journal of Pure and Applied Sciences. 1998;**1**(1):17-23

[41] Olsnes S. The history of ricin, abrin and related toxins. Toxicon. 2004;**44**(4):361-370

[42] Audi J, Belson M, Patel M, Schier J, Osterloh J. Ricin poisoning: A comprehensive review. The Journal of the American Medical Association. 2005;**294**(18):2342-2351

[43] CFSPH. The Center for Food Security and Public Health. Iowa State University; 2004. Available from: http://www.cfsph.iastate.edu/Factsheets/pdfs/ricin.pdf, https://pubmed.ncbi.nlm.nih.gov/16278363/ [Accessed: 01 July 2017]

[44] Soto-Blanco B, Sinhorini IL, Gorniak SL, Schumaher-Henrique B. *Ricinus communis* cake poisoning in a dog. Veterinary and Human Toxicology. 2002;**44**(3):155-156

[45] Islam T, Bakshi H, Sam S, Sharma E, Hameed B, Rathore B, et al. Assessment of antibacterial potential of leaves of *Ricinus communis* against pathogenic and dermatophytic bacteria. International Journal of Pharmaceutical Research and Development. 2010;**1**(12):1-7

[46] Verma SK, Yousuf SAJAD, Singh SK, Prasad GBKS, Dua VK, Mathur A. Antimicrobial potential of roots of Ricinus communis against pathogenic microorganisms. International Journal of Pharmacy and Biological Sciences. 2011;**1**(2):545-548

[47] Oloyede GK. Antioxidant activities of methyl ricinoleate and ricinoleic acid dominated *Ricinus communis* seeds extract using lipid peroxidation and free radical scavenging methods. Research Journal of Medicinal Plant. 2012;**6**(7):511-520

[48] Singh RK, Gupta MK, Katiyar D, Srivastava A, Singh P. In-vitro antioxidant activity of the successive extracts of *Ricinus communis* stems. International Journal of Pharmaceutical Sciences and Research. 2010;**1**(8):100-103

[49] Gupta MK, Sharma PK, Ansari SH. In-vitro antioxidant activity of the successive extracts of Ricinus communis leaves. International Journal of Plant Sciences. 2006;**1**(2):229-231

[50] Rakesh MR, Kabra MP, Rajkumar VS. Evaluation of antiulcer activity of castor oil in rats. International Journal of Research in Ayurveda and Pharmacy (IJRAP). 2011;**2**(4):1349-1353

[51] Shokeen P, Anand P, Murali YK, Tandon V. Antidiabetic activity of 50% ethanolic extract of *Ricinus communis* and its purified fractions. Food and Chemical Toxicology. 2008;**46**(11):3458-3466

[52] Ilavarasan R, Mallika M, Venkataraman S. Anti-inflammatory and free radical scavenging activity of *Ricinus communis* root extract. Journal of Ethnopharmacology. 2006;**103**(3):478-480

[53] Almeida RN, Navarro DS, Barbosa-Filho JM. Plants with central analgesic activity. Phytomedicine. 2001;**8**(4):310-322

[54] Ferraz AC, Angelucci MEM, Da Costa ML, Batista IR, De Oliveira BH, Da Cunha C. Pharmacological evaluation of ricinine, a central nervous system stimulant isolated from *Ricinus communis*. Pharmacology Biochemistry and Behavior. 1999;**63**(3):367-375

[55] Ladda PL, Magdum CS. Evaluation of anti-tubercular activity of *Ricinus communis* Linn. By proportion, nra and bact/alert methods. International Journal of Pharmacy and Pharmaceutical Sciences. 2012;**4**(3):474-478

[56] Sany UM, Sule MI. Anti-fertility activity of methanol extracts of three different seed varieties of *Ricinus communis* Linn (Euphorbiaceae). Nigerian Journal of Pharmaceutical Sciences. 2007;**6**(2):78-83

[57] Sandhyakumary K, Bobby RG, Indira M. Antifertility effects of *Ricinus communis* (Linn) on rats. Phytotherapy Research. 2003;**17**(5):508-511

[58] Lombard S, Helmy ME, Piéroni G. Lipolytic activity of ricin from *Ricinus sanguineus* and *Ricinus communis* on neutral lipids. Biochemical Journal. 2001;**358**(3):773-781

[59] Natu MV, Suraj A, Agarwal SL, Agarwal S. Protective effect of *Ricinus communis* leaves in experimental liver injury. Indian Journal of Pharmacology. 1977;**9**(4):265

[60] Shukla B, Visen PKS, Patnaik GK, Kapoor NK, Dhawan BN. Hepatoprotective effect of an active constituent isolated from the leaves of *Ricinus communis* Linn. Drug Development Research. 1992;**26**(2):183-193

[61] Visen PKS, Shukla B, Patnaik GK, Tripathi SC, Kulshreshtha DK, Srimal RC, et al. Hepatoprotective activity of *Ricinus communis* leaves. International Journal of Pharmacognosy. 1992;**30**(4):241-250

[62] Jena J, Gupta AK. *Ricinus communis* Linn: A phytopharmacological review. International Journal of Pharmacy and Pharmaceutical Sciences. 2012;**4**(4):25-29

[63] Prasad MK, Rachhadiya RM, Shete RV. Pharmacological investigation on the wound healing effects of castor oil in rats. International Journal of Universal Pharmacy and Life Sciences. 2011;**1**(1):1-9

[64] Taur DJ, Patil RY. Antiasthmatic activity of *Ricinus communis* L. roots. Asian Pacific Journal of Tropical Biomedicine. 2011;**1**(1):S13-S16

[65] Kumar A, Singh V, Ghosh S. An experimental evaluation of in vitro immunomodulatory activity of isolated compound of *Ricinus communis* on human neutrophils. International Journal of Green Pharmacy. 2011;**5**(3):201

[66] Beloti MM, Hiraki K, Barros VM, Rosa AL. Effect of the chemical composition of *Ricinus communis* polyurethane on rat bone marrow cell attachment, proliferation, and differentiation. Journal of

Biomedical Materials Research Part A. 2003;**64**(1):171-176

[67] Darmanin S, Wismayer PS, Podesta MTC, Micallef MJ, Buhagiar JA. Phytochemistry. An extract from *Ricinus communis* L. leaves possesses cytotoxic properties and induces apoptosis in SK-MEL-28 human melanoma cells. Natural Product Research. 2009;**23**(6):561-571

[68] Beloti MM, de Oliveira PT, Tagliani MM, Rosa AL. Bone cell responses to the composite of Ricinus communis polyurethane and alkaline phosphatase. Journal of Biomedical Materials Research Part A. 2008;**84**(2):435-441

[69] Kamalakar K, Mahesh G, Prasad RB, Karuna MS. A novel methodology for the synthesis of acyloxy castor polyol esters: Low pour point lubricant base stocks. Journal of Oleo Science. 2015;**64**(12):1283-1295

[70] Heinz PB. Practical Lubrication for Industrial Facilities. Fairmont Press; 2009. Available from: https://pubmed.ncbi.nlm.nih.gov/26582154/

[71] Singh AK. Castor oil-based lubricant reduces smoke emission in two-stroke engines. Industrial Crops and Products. 2011;**33**(2):287-295

[72] Gainer GC, Luck RM. Modified Castor Oil Lubricant for Refrigerator Systems Employing Halocarbon Refrigerants. Washington, DC, U.S; 1979. Available from: https://doi.org/10.1016/j.indcrop.2010.12.014

[73] Rudnick LR. Synthetics, Mineral Oils, and Bio-Based Lubricants: Chemistry and Technology. CRC Press; 2013. Available from: https://patents.google.com/patent/CA1098893A/en

[74] Musa U, Aberuagba F. Characteristics of a typical Nigerian *Jatropha curcas* oil seeds for biodiesel production. Research Journal of Chemical Sciences. 2012;**2**(10):7-12

[75] Salihu BZ, Gana AK, Apuyor BO. Castor oil plant (*Ricinus communis* L.): Botany, ecology and uses. International Journal of Science and Research. 2014;**3**(5):1333-1341

[76] Mutlu H, Meier MA. Castor oil as a renewable resource for the chemical industry. European Journal of Lipid Science and Technology. 2010;**112**(1):10-30

[77] Yenwo GM, Manson JA, Pulido J, Sperling LH, Conde A, Devia N. Castor-oil-based interpenetrating polymer networks: Synthesis and characterization. Journal of Applied Polymer Science. 1977;**21**(6):1531-1541

[78] Raymond MP, Bui VT. Epoxy/castor oil graft interpenetrating polymer networks. Journal of Applied Polymer Science. 1998;**70**(9):1649-1659

[79] Dave VJ, Patel HS. Synthesis and characterization of interpenetrating polymer networks from transesterified castor oil based polyurethane and polystyrene. Journal of Saudi Chemical Society. 2017;**21**(1):18-24. Available from: https://doi.org/10.1016/j.jscs.2013.08.001

[80] Chen S, Wang Q, Wang T. Hydroxy-terminated liquid nitrile rubber modified castor oil based polyurethane/epoxy IPN composites: Damping, thermal and mechanical properties. Polymer Testing. 2011;**30**(7):726-731

[81] Kunduru KR, Basu A, Haim Zada M, Domb AJ. Castor oil-based biodegradable polyesters. Biomacromolecules. 2015;**16**(9):2572-2587

[82] Maisonneuve L, Lebarbé T, Grau E, Cramail H. Structure–properties relationship of fatty acid-based thermoplastics as synthetic polymer

mimics. Polymer Chemistry. 2013;**4**(22):5472-5517

[83] Vilela C, Sousa AF, Fonseca AC, Serra AC, Coelho JF, Freire CS, et al. The quest for sustainable polyesters—Insights into the future. Polymer Chemistry. 2014;**5**(9):3119-3141

[84] Dwivedi MC, Sapre S. Total vegetable-oil based greases prepared from castor oil. Lubrication Science. 2002;**19**(3):229-241

[85] Grummitt O, Marsh D. Alternative methods for dehydrating castor oil. Journal of the American Oil Chemists Society. 1953;**30**(1):21-25

[86] Trevino AS, Trumbo DL. Acetoacetylated castor oil in coatings applications. Progress in Organic Coatings. 2002;**44**(1):49-54

[87] Thakur S, Karak N. Castor oil-based hyperbranched polyurethanes as advanced surface coating materials. Progress in Organic Coatings. 2013;**76**(1):157-164

[88] Allauddin S, Narayan R, Raju KVSN. Synthesis and properties of alkoxysilane castor oil and their polyurethane/urea–silica hybrid coating films. ACS Sustainable Chemistry & Engineering. 2013;**1**(8):910-918

[89] Lima RL, Severino LS, Sampaio LR, Sofiatti V, Gomes JA, Beltrão NE. Blends of castor meal and castor husks for optimized use as organic fertilizer. Industrial Crops and Products. 2011;**33**(2):364-368

[90] Ladda PL, Kamthane RB. *Ricinus communis* (Castor): An overview. International Journal of Research in Pharmacology & Pharmacotherapeutics. 2014;**3**(2):136-144

[91] Akpan UG, Jimoh A, Mohammed AD. Extraction, characterization and modification of castor seed oil. Leonardo Journal of Sciences. 2006;**8**:43-52

[92] Roces SA, Tan R, Da Cruz FJT, Gong SC, Veracruz RK. Methanolysis of Jatropha oil using conventional heating. Asean Journal of Chemical Engineering Research. 2011;**11**(1)

[93] Nakarmi A, Joshi S. A study on Castor oil and its conversion into biodiesel by transesterification method. Nepal Journal of Science and Technology. 2015;**15**(1):45-52

[94] Abdullah BM, Salimon J. Epoxidation of vegetable oils and fatty acids: Catalysts, methods and advantages. Journal of Applied Sciences. 2010;**10**(15):1545-1553

[95] Campbell SJ, Nakayama N, Unger EH. Chemical degumming of crude vegetable oils. United Oilseed Products Ltd. Canadian Patent. 1983:**1**(157):883

[96] Okullo AA, Temu AK, Ogwok P, Ntalikwa JW. Physico-chemical properties of biodiesel from jatropha and castor oils. International Journal of Renewable Energy Research. 2012;**2**(1):47-52

[97] Bhosle BM, Subramanian R. New approaches in deacidification of edible oils—A review. Journal of Food Engineering. 2005;**69**(4):481-494

[98] Conceição MM, Dantas MB, Rosenhaim R, Fernandes VJ, Santos IM, Souza AG. Evaluation of the oxidative induction time of the ethilic castor biodiesel. Journal of Thermal Analysis and Calorimetry. 2009;**97**(2):643

[99] Brigham RD. Castor: Return of an old crop. In: New Crops. New York: Wiley; 1993. pp. 380-383

[100] Kheang LS, Foon CS, May CY, Ngan MA. A study of residual oils recovered from spent bleaching earth:

Their characteristics and applications. American Journal of Applied Sciences. 2006;**3**(10):2063-2067

[101] Dumont MJ, Narine SS. Soapstock and deodorizer distillates from North American vegetable oils: Review on their characterization, extraction and utilization. Food Research International. 2007;**40**(8):957-974

[102] Cvengros J. Physical refining of edible oils. Journal of the American Oil Chemists' Society. 1995;**72**(10):1193-1196

[103] Naughton FC, Duneczky F, Swenson CR, Kroplinski T, Cooperman MA. Kirk-Othmer Encyclopedia of Chemical Technology. Vol. 5. New York: John Wiley & Sons, Inc; 1979. p. 1

[104] Harland SC. The genetics of *Ricinus communis* L. Bibliotheca Genetica. 1928;**4**:171-178

[105] Zimmerman LH. Castorbeans: A new oil crop for mechanized production. Advances in Agronomy. 1959;**10**:257-288

[106] Shifriss O. Sex instability in Ricinus. Genetics. 1956;**41**(2):265

[107] Ankineedu G, Rao N. Development of pistillate castor. Indian Journal of Genetics and Plant Breeding. 1973;**33**(3):416-422

[108] Hooks JA, Williams JH, Gardner CO. Estimates of heterosis from a diallel cross of inbred lines of castors, *Ricinus communis* L. Crop Science. 1971;**11**(5):651-655

[109] Prasad MVR, Rana BS. Inheritance of yield and its components in castor. ndian Journal of Genetics and Plant Breeding. 1984;**44**:538-543

[110] Giriraj K, Mensinkai SW, Sindagi SS. Components of genetic variation for yield and its attributes in 6 × 6 diallel crosses of castor (*Ricinus communis* L.) (India). Indian Journal of Agricultural Sciences. 1974;**44**:132-136

[111] Solanki SS, Joshi P. Combining ability analysis over environments of diverse pistillate and male parents for seed yield and other traits in castor (*Ricinus communis* L.). Indian Journal of Genetics and Plant Breeding. 2000;**60**(2):201-212

[112] Solanki SS, Joshi P, Gupta D, Deora VS. Gene effects for yield contributing characters in castor by generation mean analysis. Journal of Oilseeds Research. 2003;**20**:217-219

[113] Anjani K, Reddy AVP. Extra-early maturing genepool in castor (*Ricinus communis*). Journal of Oilseeds Research. 2003;**20**:213-216

[114] Sujatha M, Reddy TP, Mahasi MJ. Role of biotechnological interventions in the improvement of castor (Ricinus communis L.) and Jatropha curcas L. Biotechnology Advances. 2008;**26**(5):424-435

[115] Kolte SJ. Castor: Diseases and Crop Improvement. Delhi, India: Shipra Publications; 1995

[116] Dange SRS, Desal AG, Patel SI. Diseases of castor. In: Diseases of Oilseed Crops. New Delhi: Indus Publishing Company; 2005. pp. 211-234

[117] Soares DJ. The gray mold of castor bean: A review. In: Plant Pathology. Rijeka, Croatia: InTech Publisher; 2012

[118] Rajani VV, Parakhia AM. Management of root rot disease (*Macrophomina phaseolina*) of castor (*Ricinus communis*) with soil amendments and biocontrol agents. Journal of Mycology and Plant Pathology. 2009;**39**(2):290

[119] Basappa H. Validation of integrated pest management modules for castor

(*Ricinus communis*) in Andhra Pradesh. Indian Journal of Agricultural Sciences. 2007;**77**(6):357-362

[120] Nóbrega MDM, Andrade FD, Santos JD, Leite EJ, Azevedo DD, Lima EF. Germoplasma. O agronegócio da mamona no Brasil. Brasília: Embrapa Algodão; 2001. pp. 257-281

[121] Ribeiro LDP, Costa EC. Occurrence of *Erinnyis ello* and *Spodoptera marima* in castor bean plantation in Rio Grande do Sul State, Brazil. Ciência Rural. 2008;**38**(8):2351-2353

[122] Balint GA. Ricin: The toxic protein of castor oil seeds. Toxicology. 1974;**2**(1):77-102

[123] Greenfield RA, Slater LN, Bronze MS, Brown BR, Jackson R, Iandolo JJ, et al. Microbiological, biological, and chemical weapons of warfare and terrorism. The American Journal of the Medical Sciences. 2002;**323**(6):326-340

[124] Horton J, Williams MA. A cooker-extruder for deallergenation of castor bean meal. Journal of the American Oil Chemists' Society. 1989;**66**(2):227-231

[125] Anandan S, Kumar GA, Ghosh J, Ramachandra KS. Effect of different physical and chemical treatments on detoxification of ricin in castor cake. Animal Feed Science and Technology. 2005;**120**(1):159-168

[126] Barnes DJ, Baldwin BS, Braasch DA. Degradation of ricin in castor seed meal by temperature and chemical treatment. Industrial Crops and Products. 2009;**29**(2):509-515

[127] Diniz LL, Filho VS, Campos JMS, Valadares R, Da Silva LD, Monnerat J, et al. Effects of castor meal on the growth performance and carcass characteristics of beef cattle. Asian-Australasian Journal of Animal Sciences. 2010;**23**(10):1308-1318. Available from: https://doi.org/10.5713/ajas.2010.10041

[128] McKeon TA. Castor (*Ricinus communis* L.). In: McKeon T, Hayes D, Hildebrand D, Weselake R, editors. Industrial Oil Crops. 2016. PP. 75-112. Available from: https://doi.org/10.1016/B978-1-893997-98-1.00004-X

[129] Baldanzi M, Fambrini M, Pugliesi C. Redesign of the castorbean plant body plan for optimal combine harvesting. Annals of Applied Biology. 2003;**142**(3):299-306

[130] Doan LG. Ricin: Mechanism of toxicity, clinical manifestations, and vaccine development. A review. Journal of Toxicology - Clinical Toxicology. 2004;**42**(2):201-208

Chapter 5

Dangerous Risk Factors to be Considered for Proper Management of Agroecosystems

Carmine Massarelli, Claudia Campanale and Vito Felice Uricchio

Abstract

Our work aims to identify the main risks existing in the agroecosystems of southern Italy, providing, at the same time, information about innovative and fast methodologies. The goal is to understand the magnitude of the phenomena that could compromise them if no action is taken for water and soil matrices. Regarding the former we will consider plant protection product residues in water bodies and the importance of agroecosystems as source of microplastic pollution and their role as a vector of pollutants; regarding the latter, we will present a rapid and low-cost methodology to detect asbestos-containing materials and significantly transformed areas. Furthermore, indications are provided on how to implement effective monitoring plans in order to certainly identify the problem affecting one or more matrices and provide practical instructions to the administrators to implement the appropriate remediation strategies.

Keywords: remote sensing, microplastic pollution, asbestos, drones, data integration, GIS

1. Introduction

An agroecosystem is an anthropogenic ecosystem which constitutes the basic unit of study in agroecology; it can be defined as a spatial and functionally coherent unit in which different agricultural activities take part that includes living components (few species) and their interactions [1].

Agroecosystems provide products that can be evaluated in economic terms, following agronomic interventions on land and many biological factors.

Although the definition makes us think exclusively of the territory affected by human activity, an agroecosystem is not strictly identifiable with these areas (e.g. the farm), but instead, it includes the regions that are impacted by these activities. These regions are identifiable by changes to the species complexity (simpler species composition), energy flows (aimed at higher productivity) and nutrient balance because the cycles of a few elements are considered often associated with high nutrient input, much of which leads to connected ecosystem eutrophication [2].

For correct management of the agroecosystems, we can not ignore the information deriving from health status of environmental matrices (soil and water) assessable by fast identification of the dangers that can cause contamination to be able to

 IntechOpen

suggest possible remediation and bioremediation solutions and many other innovative technologies to improve their functionality.

The presence of harmful or dangerous substances released without any control can become a dangerous source of pollution. Many areas of the Apulia region generally, in southern Italy, are subjected to this type of phenomena. Land monitoring would be carried out in a very long time and would require significant financial resources and considerable effort if done by conventional methods.

The work has been focused on the development of an integrated methodology with a defined and high reliability capable of identifying the presence of dangerous sources of pollution for agroecosystems: asbestos, illegal waste burial for soil and microplastic pollution.

1.1 Asbestos

Asbestos are naturally widespread minerals belonging to inosilicate group (amphibole series) and the phyllosilicate group (serpentine series). At first they have been widely used for their intrinsic resistance to heat and its fibrous structure (suitable for buildings and fireproof fabrics), but later it was ascertained that it is harmful to health at the point of being prohibited in many countries. Dust-containing asbestos fibers are responsible for serious diseases such as asbestosis, pleural mesothelioma and lung cancer.

Before discovering its danger to humans and the environment, asbestos was widely used in several applications; the main one above all is for insulation of buildings and rooftops in the form of composite fiber cement (also known "Eternit"). Given its proven hazard to human health, there are numerous scientific studies in which its presence is investigated through remote sensing technics and sensors, e.g. multispectral visible and infrared imaging spectrometer (MIVIS) [3] or visible and thermal Landsat images [4].

1.2 Illegal waste burial

A severe threat to human health and agroecosystems is represented by illegal waste burial and ground dumping of polluted sludge, especially if it concerns hazardous industrial waste. These illegal activities mainly take place in large areas that undergo large transformation processes in a short time like quarry areas and landfills, even if it is licensed. Unfortunately, since we have witnessed the countryside abandoning phenomenon, such phenomena are increasing because a lot of sites remain increasingly unattended and uncontrolled. Sometimes, entrepreneurs may receive pressure from criminal organizations to buy hectares of farmland in order to have more and more areas to grow the eco-mafia business. In these areas, we have not gone looking for changes such as cultural growth, different photosynthetic activities or other distinctive elements. These elements find more applications in land use-land cover classification, but only in those aspects related to surface soil change, alterations or reshuffle to hide illegal waste dumping that affect the characteristics of vegetation, soil texture and moisture content.

1.3 Microplastic pollution

Both in Europe and many parts of the world, the contamination of soils by plastic is growing. Plastic waste from the terrestrial environment constitutes about 80% of all plastic debris found in the marine environment, representing a source of pollution not only for the seas but also for inland waters and soils, even if the phenomenon is currently less known and studied. The plastic in direct contact with the soils mainly comes from various widespread practices adopted in agriculture, such as mulching [5], which uses black plastic sheeting positioned on the ground to prevent moisture loss and growth of weeds and to retain heat in spring. Plastic undergoes a series of physical, chemical and biological reactions due to the effects of UV radiation, atmospheric agents and the action of the organisms that inhabit the soils. This cause its embrittlement which leads to the degradation of polymeric waste in smaller and smaller fragments going from macroplastic (>25 mm) to mesoplastic (5–25 mm) and from mesoplastic to microplastic (<5 mm) that creep into agricultural soils and terrestrial ecosystems.

Besides, darker-colored plastic materials, such as mulch sheets and irrigation materials used in agriculture, absorb more sunlight, and the consequent increase in temperature leads to faster decomposition and higher production of meso- and microplastics by fragmentation.

Microplastics can reach terrestrial environments and in particular agroecosystem through various input sources such as soil mulching, the use of compost in agricultural soils, irrigation, rainwater and atmospheric fallout.

Moreover, microplastics, due to their hydrophobic and microscopic size that influence a high S/V ratio, represent ideal vectors for the adsorption of environmental pollutants, especially persistent organic contaminants (chlorinated organ pesticides, PCBs, IPAs) [6], up to several orders of magnitude higher than those present in the surrounding environment [7]. Therefore, agroecosystems, already sensitive to chemical pollution mainly represented by the waste of plant protection products used in agriculture and more exposed to the presence of macro-, meso- and micro-plastics due to the practices used in agriculture, represent excellent tools for studying and investigating the interaction between micropollutants and plastics.

In the absence of global measures to regulate the use of plastic in contact with soils, integrated monitoring of synthetic polymers and adsorbed micropollutants is therefore of fundamental importance in order to investigate the problem of plastics and microplastics in soils, currently almost totally unknown.

2. Materials and methods

2.1 Study areas

For the study areas, we consider three different territories which are all agroecosystems but with different risk factors:

1. The surrounding territory of the municipality of Brindisi in Southern Italy (**Figure 1**). The perimeter is about 22.000 km, the area is 31.00 km^2, and it is characterized by the presence of different farming practices, greenhouses, small structures built primarily for depository purposes, roads (mostly unpaved) and a lot of caves.

2. This study area is particularly affected by abandonment phenomena. It is called Gravina (ravine) of Leucaspide, its extension is approximately 5 ha in Statte municipality (Taranto Province), and it also represents the greatest example of karst in the Taranto area (**Figure 2**).

3. The Ofanto river, the most important river in the Apulia region (Italy) for length, area and abundance of water. Its source is at 715 m above sea level, in the province of Avellino, and it crosses part of Campania and Basilicata regions flowing then into the Adriatic Sea, between the towns of Barletta and

Margherita di Savoia. The shape of the river is trapezoidal with a surface of 2790 sq. km and a mean altitude of 450 m. The length of the main boom is about 165 km, the average annual inflow of 720 mm, and the mean annual temperature is just over 14°C [9]. The region of Ofanto river covers about 88.700 hectares, of which 8% are natural (6.800 ha). The predominant agricultural areas include nonirrigated (30.000 ha) and irrigated (14.000 ha) arable land, which, in total, represent 50% of the territory. In the floodplain of the river, vineyards prevail above all (18.400 ha), followed by the olive groves (14.100 ha) and the orchards (1.600 ha) (**Figure 3**). These permanent crops make up 39% of the area. Lastly, the urbanized district covers 3% (2700 ha) [10].

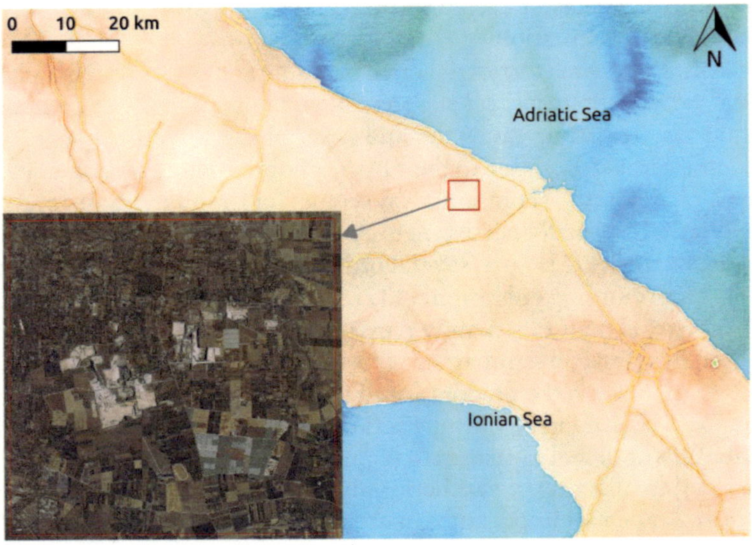

Figure 1.
Study area. Map tiles by stamen design [8], under a creative commons attribution (CC BY 3.0) license. Data by OpenStreetMap, under CC BY SA.

Figure 2.
The perimeter of Gravina of Leucaspide (south of Italy) in red. Map tiles by stamen design [8], under a CC BY 3.0 license. Data by OpenStreetMap, under CC BY SA.

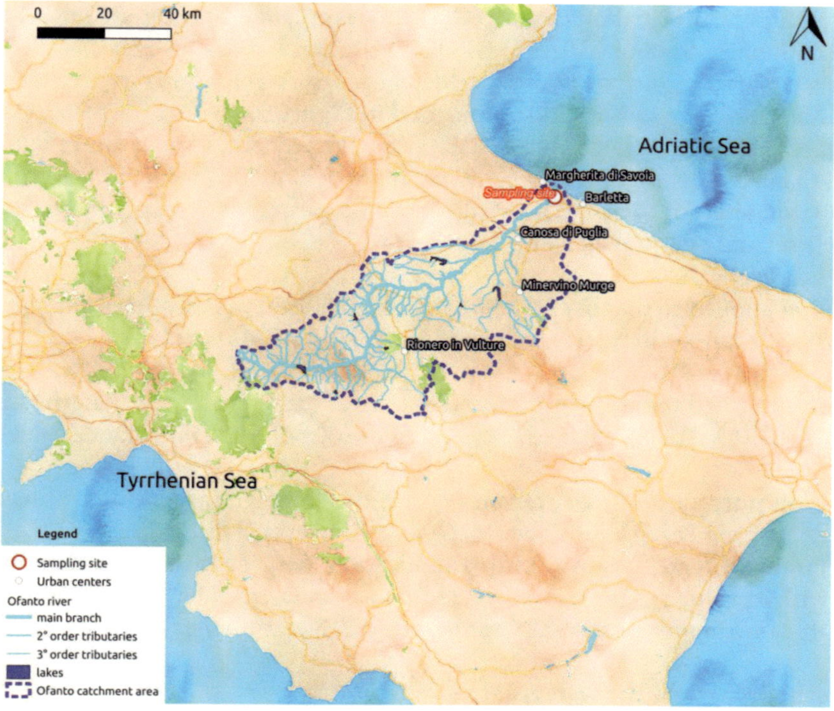

Figure 3.
The study area of Ofanto river. The red fringe represents the river catchment area. Data source: EEA (2018). Map source: Map tile sets are © stamen design, under a CC BY 3.0 license.

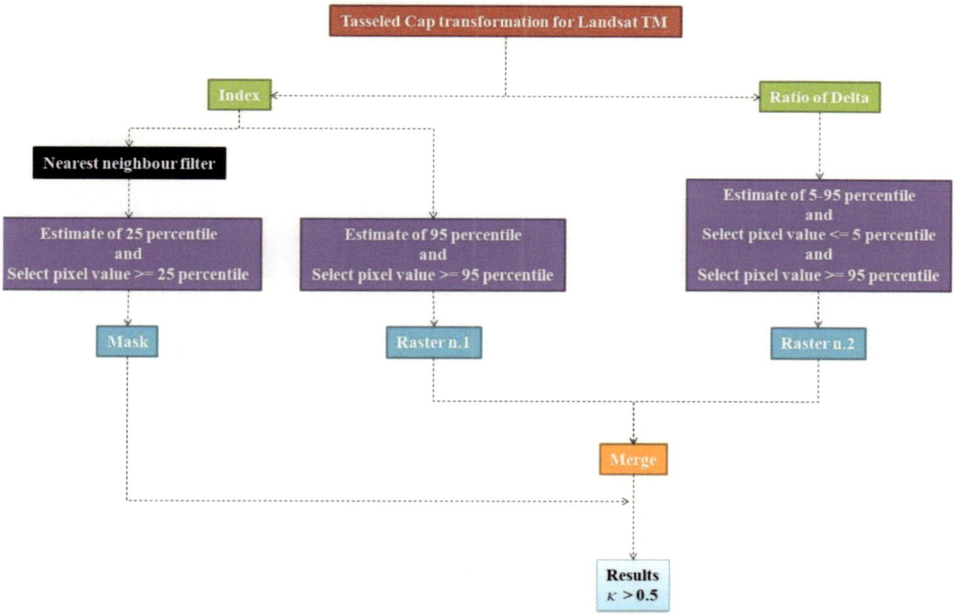

Figure 4.
Operational workflow for illegal waste burial detection.

2.2 Methodology for the soil matrix

For synthesis and completeness reasons, the whole methodology for illegal waste burial detection is shown in **Figure 4**. All the phases will be analyzed in detail in the following sections. All the images processing have been performed with the Geographic Information System (GIS) software called Geographic Resources Analysis Support System (GRASS) GIS [11, 12] using Landsat images.

The whole methodology adopted to detect asbestos-containing materials (data processing and used sensors) is summarized in **Figure 5**.

Image processing was done in GRASS GIS environment. First of all we found it useful to make a distinction between two ranges of bands: 375–702 nm and from 717 to 1030 nm; later for each one we applied four filters because we noticed that it was possible separate natural elements by asbestos:

- Filter applied to the first 24 of CASI raster bands in the wavelength range 375–702 nm to eliminate background value (no ACM included)

- Filter applied to the second 24 of CASI raster bands in the wavelength range 717–1030 nm to eliminate the emerging rock from ACM reflectance values

For this purpose it was written an algorithm able assigns a value of 1, otherwise 0 to every pixel in each band if not filtered or vice versa if filtered.

Summing all pixel values after this application, we get a raster with only one band with its value ranging from 0 to 24 that we could consider equivalent to 0–100% of fitting pixels in regard to the spectral signature measures.

2.3 Methodology for the water matrix

To monitor the trend of microplastic abundances over a long period, seven seasonal sampling campaigns were carried out. Microplastic samples were collected form river surface water during February, April, October and December 2017, May and December 2018 and April 2019, all of them collected from the same point located at 6 km from the Ofanto river mouth following the sampling strategy reported here [13].

All water samples picked up during each campaign were preserved in glass containers and processed once in laboratory to extract microplastics following

	CASI sensor on airplane	RGB camera on drone	FIELDSPEC Iperspectral sensor	GROUND TRUTH
Preparation for data acquisition	Set configuration relative to the acquisition bands, flight planning	Flight planning	-	-
Data acquisition	4 acquisitions	~100 RGB images	4 sets of asphalt spectra measurements	GPS differential point acquisition over areas containing asbestos
Data pre-processing	Geometric, radiometric and atmospheric correction	Images mosaicking, asbestos identification	Averaging of spectra removing atmospheric influence	Ground truth raster building of known asbestos areas
Data processing	Classifications methods with mask generation and filters applications	-	-	-
Results and statistical analysis				

Figure 5.
Operational workflow for asbestos-containing materials detection.

the method reported by [13]. After processing, all the microplastics were visually identified under a 40× digital microscope (Keyance VH-Z 100 UR). All plastic microparticles detected were counted, photographed, enumerated and categorized based on color (black, transparent and colored) and morphology (fragments, flakes, pellets, lines, fibers, films and foams) (**Figure 6**).

Microplastic concentrations were expressed as mean values (±DEV. ST.) of six replicates for each campaign. Concentrations were indicated as the number of particles per cubic (p/m^3).

Simple regression and Spearman's non-parametric correlation coefficient were used to test significant relations between the concentration of microplastics and the water level of the river for each monitoring campaign. Spearman correlation test was performed by Statgraphics Centurion Software.

2.4 Data integration

The main objective is to evaluate the level of degradation and possible contamination of environment through the combination of different types of investigations in soil and water environmental matrices.

This goal can only be achieved by implementing a data integration procedure to make them comparable and so that each type of investigation, representative factor of the state of a matrix, can be used with others in order to identify compromised environments.

The more representative factors we have, the more reliable the result will be.

In particular our goal is to locate the same zoning areas subject to a different state of degradation.

From the greater knowledge of the environmental component investigated, considerable progress can be derived for the identification and implementation of technologies useful for monitoring and/or for the safety and remediation of polluted matrices.

First of all it is necessary to create data matrices deriving from monitoring and sampling activities. Having the different variables different units of measure (sometimes even of many orders of magnitude), a matrix normalization was carried out [14]: the standard deviation of each variable was first calculated, and each value was subsequently divided for the latter, thus obtaining dimensionless matrices. Later the cluster analysis was carried out. It is a sector of multivariate analysis that

Figure 6.
Microplastic assortment collected by Ofanto river subdivided by morphology as: Fragments and flakes (broken pieces of larger debris), pellets (preproduction pellets), lines and fibers (particles of fishing line and nets and fibers from synthetic textiles) and foams (foam cups, takeout containers, packaging).

groups numerous techniques (k-means, fuzzy k-means, hierarchical cluster, etc.) that allow to group monitoring units (e.g. sampling points) that present characters (the different variables) of similarity between them [15].

Two clustering methods based on very different theoretical approaches have been applied in order to achieve a robust result: hierarchical clustering and k-means [14].

The choice to opt for these methods for the classification of the points investigated was based, above all, on the purpose of the work, i.e. defining homogeneous areas with a priority of intervention (further environmental investigations with more or less immediate action). In this regard, to avoid that, points with similar characteristics are distributed randomly over the territory; we wanted to consider, among the variables, the geographic coordinates that act as "attractors" for the delineation of the clusters.

Through hierarchical cluster analysis and the analysis of the resulting dendrogram, it was possible to identify the optimal number of clusters. This value was equal to 3. Through the clustering procedure, the software groups the dataset into classes based on the similarity index but is not able to independently discriminate the "danger" to be attributed to each of the classes. For this purpose, the centroids of the three classes were analyzed, and it was, therefore, possible to define the "most impacted" class, the "average impacted" class and the "least impact" class that is made based on the interpretation of the distance from the centroids for the dataset relating to the samples.

The results obtained through cluster analysis make it possible to identify geographic areas in which homogeneity of degradation levels is presumed.

The methodology used, starting from the data produced by the cluster analysis, is based on an approach that partially follows that used for the realization of the conceptual model of the site in contaminated areas [16].

Therefore the following steps were carried out in the GIS environment:

1. Distribution of the entire area of interest in different areas of "presumed homogeneity" through the identification of the Voronoi polygons.

2. Attribution to the various polygons of the "Alert" value coming from the cluster analysis.

3. Classification of the different polygons according to the "Alert" value.

4. Identification of an empirical threshold value of 200 Ha with which to identify areas with lower probability of certainty of the result coming from the cluster analysis (these areas would need further sampling).

5. The results of zoning identify areas in which it is necessary to implement short-term (red), medium-term (yellow) and long-term (green) interventions.

Further deductions can be made by considering the adjacency between polygons of different colors (e.g. a green polygon surrounded by red polygons has to be considered red) and assigning different threshold values according to other territorial characteristics.

3. Results and discussion

The results obtained for this purpose are considered satisfactory because it is possible to identify not only the potential sites where there may have been illegal activities but also the complexity of the highly transformed areas. This increases awareness to authorities about all transformation and impacts on our fragile agroecosystems because agricultural activities and related soil processes do not show significant changes compared to those highlighted through the application of these procedures. Although κ may seem low, it should not be overlooked in these procedures because the purpose, with only two Landsat images used in a short time, is to identify certain type of changes relatable to illegal waste disposal; increasing the number of processed images, it is reasonable to think that the accuracy of the results may increase, but on the other hand it will be necessary to invest in more resources (also storage, hardware, staff engaged in work activities). Presence of solar panels and building structures with metal roof alters the result; so, it would be appropriate to exclude them realizing an ad hoc mask to reduce false positives.

The developed methodology for remote sensing analyses the spectral behaviour of materials, highlighting and emphasizing certain features through the use of a procedure based on an if-then-else control structure. It also allows the selection of the most useful features to be combined that significantly reduces the number of false positives.

Here the results of accuracy and specificity for the filters are reported (**Figure 7**).

Results obtained correctly predict 99.87% of cases with 0.13% error for reduced ground truth area and correctly predict 99.78% of cases with 0.22% error for augmented one. We have pixels identifying ACM correctly in 29.49% times in reduced area, while 21.82% in augmented one. Results show a correctness probability of 99.9% with a 0.10% of false alarm probability in the case of reduced ground truth area (99.93 and 0.007%, respectively, for the augmented one). For the other filters that also provide high statistical indices, we have to consider that the number of significant pixels is much higher than those of filter #1. Starting from a computation

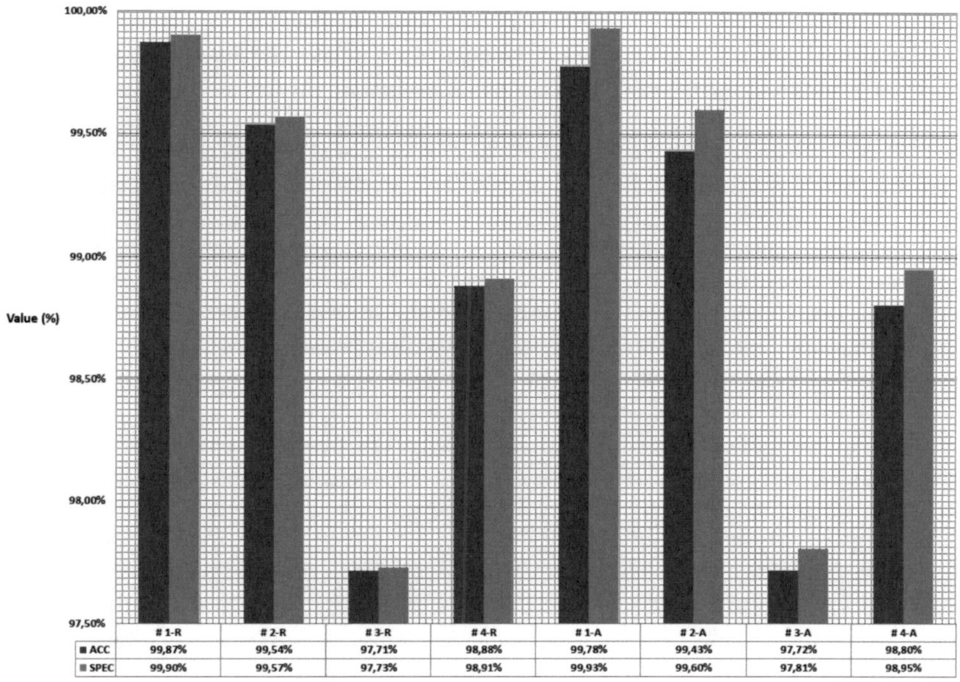

Figure 7.
Accuracy (ACC) and specificity (SPEC) values of the different filters for two different ground truth areas: Reduces (R) and augmented (A) [17].

region of 190,437 pixels, filter #1 consists only of 212 pixels, while filter #2 of 830, #3 of 4383, and #4 of 2023. This immediately translates into the lower surface to examine where ground investigations can be focused. Such considerations may also be deduced by the comparison of histogram filters considering the values of accuracy and specificity: filter #1 always shows higher values. This confirmation also means that the photointerpretation by drone is correct and can be used to supplement the investigations.

A summary of the data from the kappa analysis is shown in **Table 1**.

Considering the error matrix [19], we know that the columns represent classification depending on our knowledge and rules given in the supervised method, while the rows represent classification coming from application of the specific algorithm. We look at the negative category column; it reports that 3243 cells were correctly classified as negative; however, 550 cells were classified as positive when they were, in fact, negative. If we sum the pixels number in diagonal, we get 3869 of correctly classified pixels, if we sum the 'col sum', we get 4600 pixels representing all those considered correctly classified. So, dividing the first one by the second, the results is 0.8411, which is equivalent to 84.11%; here we have accuracy value of the classification algorithm. Observing the error statistics section of our table and considering the percentage commission column, we can see how many cells were placed into its class incorrectly: so positive pixels are confused with negative ones 46.77% of the time. This value could be considered high because represents the time that we can mistakenly attribute to the positive pixels, even if the most important thing is not to exclude true positives. This aspect given the purpose of the study is not a limiting factor as it is preferably a false alarm which indicates that the areas after a field survey are not affected by the illegal conduct, rather than a failure alarm where affected areas were not reported. The percentage omission column represents pixels placed incorrectly into other classes.

In section (d), we have the estimated kappa coefficient (κ) as a statistical value of the degree of classifications which are overlapping (more intense training of the dataset which leads to a greater agreement value of the classification results). κ takes into account the agreement of classification versus the possibility that the agreement is just from sheer chance (both classifiers are just randomly guessing the classes). If the classifiers agree on all classifications, then κ would equal to 1. If the classifiers do not agree other than what would be expected by sheer chance, then κ would equal to 0. So, in our case, κ coefficients are reported as 0.53. This could be considered a moderate agreement. A possible interpretation of κ indicates a moderate agreement for the workflow adopted, even if there is no universally agreed-upon range of values that would consider a κ coefficient to be excellent, good, moderate, weak or otherwise and is referred as common thinking. Here is one possible interpretation of κ: weak agreement (<0.20), fair agreement ($0.20 < \kappa < 0.40$), moderate agreement ($0.40 < \kappa < 0.60$), good agreement ($0.60 < \kappa < 0.80$) and excellent agreement ($0.80 < \kappa < 1.00$).

For our purposes concerning the monitoring of microplastics along the Ofanto river, we evaluated the seasonal trend of the concentration of particles hypothesizing their alleged origin. The quantitative analysis showed the presence of MPs in each sample analysed, counting a total of 164,143 microplastic particles during all the campaigns. The lowest abundances were detected in the months of October 2017 (0.93 ± 0.4 p/m^3), December 2017 (1.12 ± 0.37 p/m^3) and December 2018 (2.59 ± 0.34 p/m^3), while the highest ones in the months of May 2018 (12.56 ± 4.83 p/m^3), February (10.21 ± 4.29 p/m^3) and April 2017 (5.16 ± 1.4 p/m^3) with a maximum peak detected in April 2019 (36.05 ± 9.80 p/m^3) (**Figure 8**).

The largest number of microplastics were detected during wet periods (February 2017, May 2018 and April 2019) suggesting a presumable land-based origin from the

DOI: http://dx.doi.org/10.5772/intechopen.91824

		Procedure application		**Reference map**		
a)		Classification map		*N*	*P*	Row sum
			N	3243	181	3424
			P	550	626	1176
			Col sum	3793	807	4600
b)		Observed corrected	3869			
		Total observed	4600			
		% Observed correct	84.11			
c)		% Commission	% Omission			
	N	5.2	14.5			
	P	46.7	22.4			
d)		*k*	Kappa variance			
		0.53	0.000199			

N, number of negative pixels; P, number of positive pixels

Table 1.
Kappa analysis procedure with site-specific calculated parameters [18].

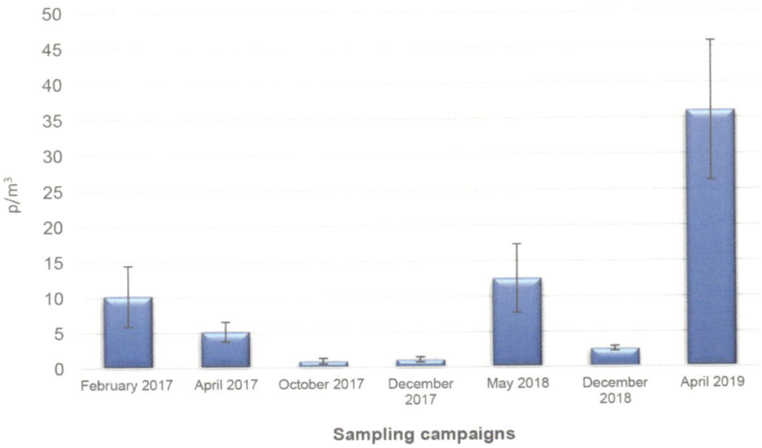

Figure 8.
Microplastic concentrations (mean value of six replicates ± dev. st.) of the sampling campaigns.

surrounding agricultural areas. As already evidenced by [20], Spearman's correlation results show a positive statistically significant correlation between the concentration of MPs and the water level ($r = 0.6583$, $p = 0.01$) of Ofanto river, indicating a marked relationship between the two factors.

It is not surprising that runoff occurring during more abundant rains can transport more microplastic debris from land to water. Indeed, it is recorded that rainy events increase the plastic concentration up to 150 times in an urban part of the Rhone river basin (France) [21, 22].

In the Los Angeles River (USA), microplastic densities were highest in samples collected in the wet season and near the surface of the water rather than samples took in the dry season and in the mid-column or near the bottom of the water column or the riverbank [23].

Others authors [24] investigated the presence of microplastic particles (fragments, foams, films and pellets/beads) in 29 Great Lakes tributaries deepening the role of hydrology in the occurrence of plastic, and they found higher concentrations during runoff events than during low-flow condition.

Different shapes and colors of microplastics were observed and quantified (**Figure 9**). Fragments and flakes were present in all samples of each campaign, constituting the most prevalent morphology identified. The mean percentage of fragments observed during campaigns was of 44%, followed by flake particles (36%) the second most abundant category counted. The percentage distribution of the other categories highlighted an almost equal distribution of lines and fibers (8% of fibers and 9% of lines) followed by pellets, foils and foams ≤3% (**Figure 9**).

Morphological information regarding microplastics is a useful tool to indicate their potential origins.

For example, flake particles, a new plastic morphology observed regularly in Ofanto river [13] in large quantities (**Figure 9**), appeared mainly of black color with an irregular shape and rough embrittled and weathered surface. The aspect of plastic particles usually depends on the fragmentation process occurred as well as the stay time in the environment [25]; breakdown of particles due to biological, chemical and weathering processes causes the degradation and erosion of microplastics through the formation of visible cracks on the plastic surface that produce a wide variety of different shapes [25].

The general irregular aspect of flakes found in Ofanto river seems to suggest an ongoing break-up process probably due to a physical and mechanical degradation

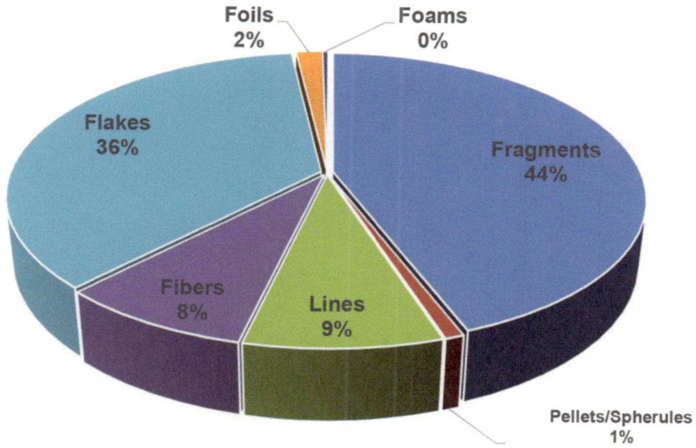

Figure 9.
Shape composition of microplastics in different campaigns, expressed as mean relative abundance in percentage.

(runoff, abrasive forces, heating/cooling, freezing/thawing, wetting/drying) that have continuously scratched their surface.

Moreover, in the presence of sunlight, all plastics undergo photo-oxidation reactions causing embrittlement and reducing the physical stress needed for fragmentation [26, 27]. Plastics exposed to a direct source of solar UV radiation, lied on beaches or soils, undergo a very efficient mechanism of degradation. Moreover, darker-colored plastic objects (e.g. black particles) would be expected to absorb more sunlight, and the resulting increase in temperature leads to more rapid decomposition. Differently, when the same plastic material is exposed to sunlight while floating in the water, degradation is severely slowed due to the fact that all different-colored plastic particles will be at the same river water temperature and decomposition rates will not vary with differential heating caused by the different colouration of microplastics [28]. Therefore, the irregular, jagged and embrittled appearance of flakes observed in the Ofanto river indicate a more advanced decomposition process with respect to transparent particles. This happens because where the plastic debris is pigmented dark, the heat build-up due to solar infrared absorption can raise its temperature even higher. The light-initiated oxidative degradation that is accelerated at higher temperatures by a factor depending on the activation energy E_a of the process, where the $E_a \sim 50$ kJ/mole, for instance, is the rate of degradation that doubles when the temperature rises by only 10°C [26, 28]. The result of this mode of oxidative degradation is a weak, brittle surface layer that develops numerous microcracks [28–31]. This degraded fragile surface is susceptible to fracture by stress induced by humidity or temperature changes as well as abrasion against surfaces [28, 32], generating particles similar to flakes identified in Ofanto river. The same degradation does not occur in plastics exposed while floating in water, suggesting, therefore, a further confirmation of the land-based origin of plastic particles in Ofanto river. The land-based origin of flake particles found in Ofanto river could be associated mainly with agricultural activities, which represent the predominant use of the land area of Ofanto valley. Plastic in direct contact with the soil comes mainly from various practices spread in agriculture, including mulching, which uses black plastic polyethylene sheets placed on the ground to prevent the loss of moisture and the growth of weeds and to retain in spring the heat and irrigation through dripping wings in black polyethylene. Recognition that microplastics (and therefore also nanoplastics) are most likely generated on beaches or riverside or inner lands underlines the importance of cleaning actions

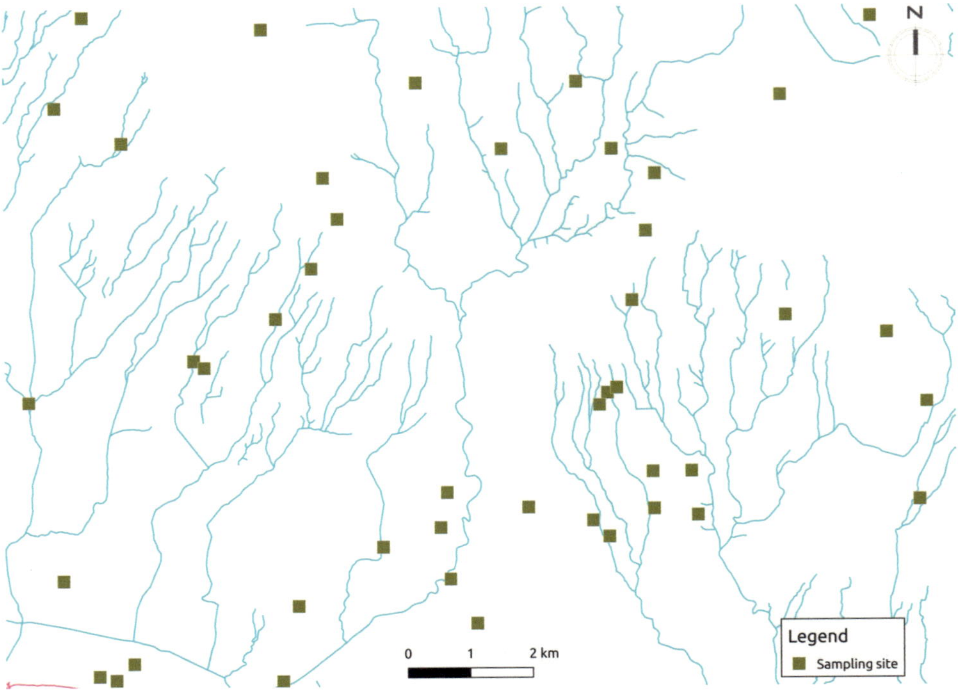

Figure 10.
Test area with rivers.

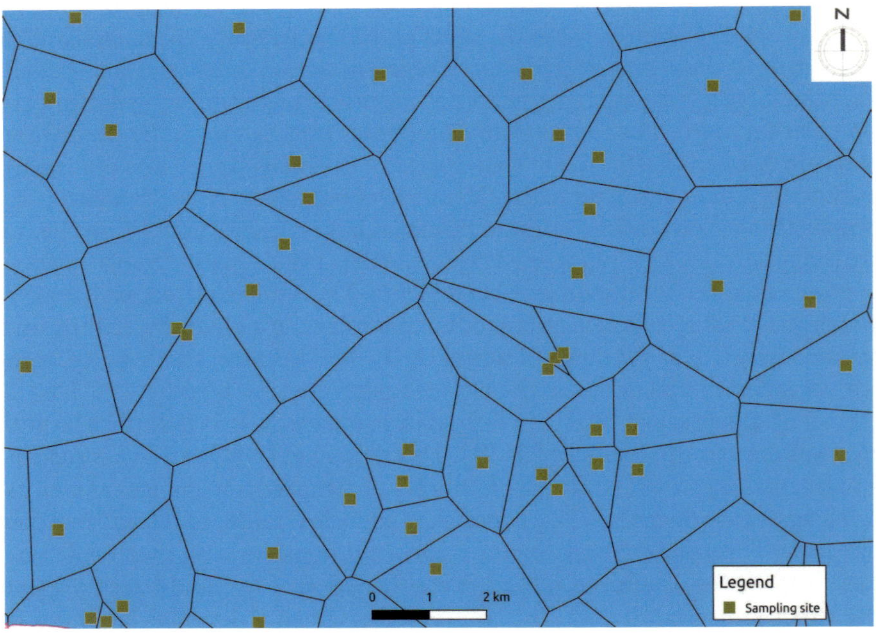

Figure 11.
Division of the test area into Voronoi polygons.

as an effective mitigation strategy contributing towards the health of the food web. The removal of larger pieces of plastic debris from beaches and riversides before these are weathered enough to be surface embrittled can have considerable value in reducing the microplastics that end up in rivers and oceans.

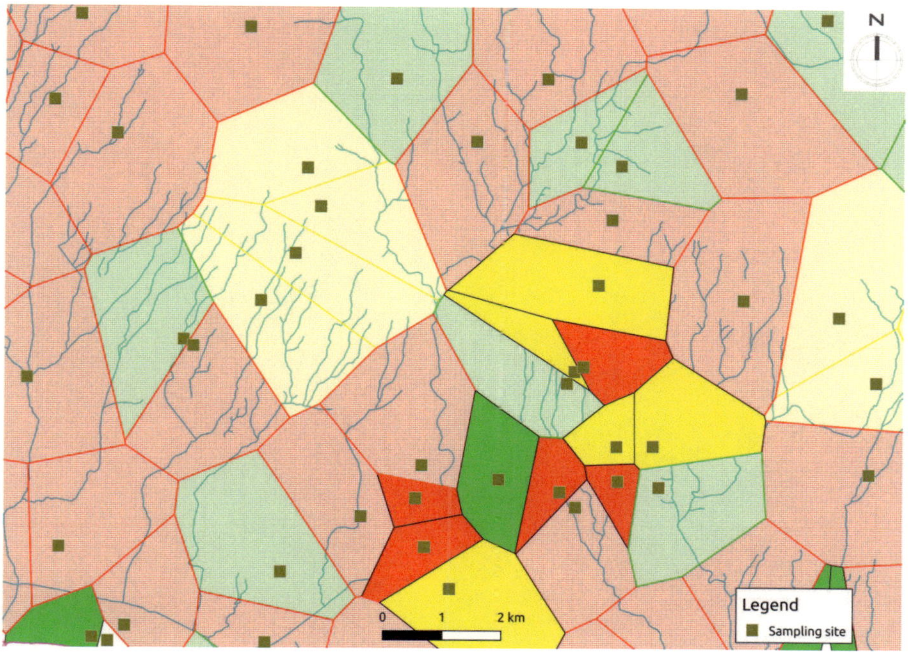

Figure 12.
Results of the classification according to the methodology proposed in the data integration paragraph.

Finally, if we consider a test area with the following sampling sites and rivers (**Figure 10**), we divide the territory with the Voronoi polygons (**Figure 11**), and then applying what is reported in the data integration paragraph, we obtain the result shown in **Figure 12**.

4. Conclusions

The results of the mapping of homogeneous areas of degradation, carried out through the cluster analysis of the data produced by various activities, express a high level of internal coherence evident also concerning the chemical–physical results only. Therefore the radiometric and chemical data are confirmed to be functional for the implementation of methodologies for the rapid identification of degradation levels.

The results of the zoning carried out through the methodological proposal described in this document indicate areas in which it is necessary to implement short-term, medium-term and long-term interventions.

Further deductions can be derived by considering the adjacency between polygons of different colors (e.g. a green polygon surrounded by red polygons is to be considered red, the proximity to inhabited centers, etc.) and attributing different threshold values according to additional territorial characteristics.

In conclusion, the validation of predictive models obtained by integrating chemical-physical analyses with radiometric data is an added value to the methodology proposed here to implement innovative, fast and economically advantageous monitoring technologies. With the proposed methodology, it will be possible to use any data source to obtain evermore precise information on the state of degradation of the environment.

Author details

Carmine Massarelli*, Claudia Campanale and Vito Felice Uricchio
Water Research Institute of Italian National Research Council, Italy

*Address all correspondence to: carmine.massarelli@ba.irsa.cnr.it

References

[1] Smit B, Waltner-Toews D, Rapport DJ, Wall E, Wichert G, Gwyn E, et al. Agroecosystem Health: Analysis and Assessment. University of Guelph, Ontario: M&T Insta-Print; 1998

[2] van de Fliert E, Braun AR. Farmer field school for integrated crop management of sweetpotato. In: Field Guides and Technical Manual. Bogor, Indonesia: International Potato Center; 1999. ISBN 92-9060-216-3

[3] Frassy F, Candiani G, Rusmini M, Maianti P, Marchesi A, Rota Nodari F, et al. Mapping Asbestos-cement roofing with Hyperspectral remote sensing over a large mountain region of the Italian Western Alps. Sensors. 2014;**14**(9):15900-15913. DOI: 10.3390/s140915900

[4] Petja BM, Twumasi YA, Tengbeh GT. The use of remote sensing to detect Asbestos mining degradation in Mafefe and Mathabatha, South Africa. Denver, CO: International geoscience and remote sensing symposium (IGARSS); 2006. pp. 1591-1593. Art no. 4241557

[5] Bläsing M, Amelung W. Plastics in soil: Analytical methods and possible sources. Science of the Total Environment. 2018;**612**(2018):422-435

[6] Hong SH, Shim WJ, Hong L. Methods of analysing chemicals associated with microplastics: A review. Analytical Methods. 2017;**9**:1361

[7] Campanale C, Massarelli C, Savino I, Locaputo V, Uricchio VF. A detailed review study on potential effects of microplastics and additives of concern on human health. International Journal of Environmental Research and Public Health. 2020;**17**:1212. DOI: 10.3390/ijerph17041212

[8] Stamen. Map Tiles Courtesy BY Stamen Design, under CC BY 3.0. Data BY OpenStreetMap, under CC BY SA. 2016. Available from: http://stamen.com/ [Accessed: 12 October 2017]

[9] Gentile F, Bisantino T, Milella P, Trisorio Liuzzi G. Relazioni tra trasporto solido e variabili idrologiche nei corsi d'acqua della Puglia Settentrionale. Periodico dell'Ordine dei Geologi della Puglia. 2007;**3-4**:61-68

[10] EEA (European Environmental Agency). Copernicus land monitoring service - corine land cover. 2018. Available from: https://www.eea.europa.eu/data-and-maps/data/copernicus-land-monitoring-service-corine.

[11] GRASS Development Team. Geographic resources analysis support system (GRASS) software, version 7.2. Open source geospatial foundation. 2017. Available from: http://grass.osgeo.org

[12] Neteler M, Bowman MH, Landa M, Metz M. GRASS GIS: A multi-purpose open source GIS. Environmental Modelling & Software. 2012;**31**(May): 124-130. DOI: 10.1016/j.envsoft.2011.11.014

[13] Campanale C, Massarelli C, Bagnuolo G, Savino I, Uricchio VF. The problem of microplastics and regulatory strategies in Italy. Springer Plastics in the Aquatic Environment - Part II: Stakeholders Role against Pollution. The Handbook of Environmental Chemistry. Switzerland: Springer Nature Switzerland AG; 2019. DOI: 10.1007/698_2019_419

[14] Delvecchio F. Statistica per l'analisi di dati multidimensionali. Padova, PD, Italy: CLEUP Editore; 2010

[15] Andemberg M. Cluster Analysis for Applications. NY, USA: New York Academic Press; 1973

[16] U.S. EPA. RAGS: volume 3 PART A-Process for Conducting Probabilistic Risk Assessment_Appendix A; 2001c

[17] Massarelli C, Matarrese R, Uricchio VF, Muolo MR, Laterza M, Ernesto L. Detection of asbestos-containing materials in agro-ecosystem by the use of airborne hyperspectral CASI-1500 sensor including the limited use of two UAVs equipped with RGB cameras. International Journal of Remote Sensing;**38, 2017**:8-10, 2135-8-2149

[18] Massarelli C. Fast detection of significantly transformed areas due to illegal waste burial with a procedure applicable to landsat images. International Journal of Remote Sensing. 2018;**39**(3):754-769

[19] Smith R, Texas A&M University – Corpus Christi. GST 105: Introduction to Remote Sensing Lab 6: Accuracy Assessment. Texas: National Information Security, Geospatial Technologies Consortium (NISGTC); 2014

[20] Campanale C, Stock F, Massarelli C, Kochleus C, Bagnuolo G, Reifferscheid G, et al. Microplastics and their possible sources: The example of Ofanto river in Southeast Italy. Environmental Pollution. 2020;**258**:113284. ISSN: 0269-7491. https://doi.org/10.1016/j.envpol.2019.113284

[21] Kooi M, Besseling E, Kroeze C, van Wezel AP, Koelmans AA. Modeling the fate and transport of plastic debris in freshwaters: Review and guidance. In: Wagner M, Lambert S, editors. Freshwater Microplastics: Emerging Environmental Contaminants? Heidelberg: Springer; 2017. DOI: 10.1007/978-3-319-61615-5_7

[22] Wagner M, Lambert S. Freshwater microplastics. In: Emerging Environmental Contaminants? The Handbook of Environmental Chemistry, Vol. 58. Cham, Switzerland: Springer Open; 2018

[23] Moore CJ, Lattin GL, Zellers AF. Quantity and type of plastic debris flowing from two urban rivers to coastal waters and beaches of Southern California. Journal of Integrated Coastal Zone Management. 2011;**11**(1):65e73

[24] Baldwin AK, Corsi SR, Mason SA. Plastic debris in 29 great lakes tributaries: Relations to watershed attributes and hydrology. Environmental Science & Technology. 2016;**50**:10377-10385

[25] Hidalgo-Ruz V, Gutow L, Thompson RC, Thiel M. Microplastics in the marine environment: A review of the methods used for identification and quantification. Environmental Science & Technology. 2012;**46**:3060-3075

[26] Shaw DG, Day RH. Colour- and form-dependent loss of plastic micro-debris from the North Pacific Ocean. Marine Pollution Bulletin. 1994;**28**:39-43

[27] van Krevelen DW. Properties of Polymers: Correlations with Chemical Structure. Amsterdam: Elsevier; 1972

[28] Andrady AL. Microplastic in the marine environment. Marine Pollution Bulletin. 2011;**62**:1596-1605

[29] Qayyum MM, White JR. Effect of stabilizers on failure mechanisms in weathered polypropylene. Polymer Degradation and Stability. 1993;**41**:163-172

[30] Blaga A, Yamasaki RS. Surface microcracking induced by weathering of polycarbonate sheet. Journal of Materials Science. 1976;**11**:1513-1520

[31] Blaga A. In: Sereda PJ, Litvin GG, editors. Durability of Building Materials and Compounds. Philadelphia: ASTM STP 691, ASTM; 1980. pp. 827-837

[32] George GA. Weathering of polymers. Metals Forum. 1995;**19**:145-161

Chapter 6

Use of Deep Eutectic Solvents in the Treatment of Agro-Industrial Lignocellulosic Wastes for Bioactive Compounds

Ayşe Ezgi Ünlü and Serpil Takaç

Abstract

Lignocellulose is the most abundant component in nature since it refers to plant material. Beyond the enormous utilization of lignocellulose by human being, unignorable amount of waste is also formed simultaneously. Agro-industrial lignocellulosic wastes can cause environmental pollutions if not processed before discharged. An innovative approach for lowering the detrimental influences of lignocellulosic wastes is to consider them as a source of useful products rather than a waste to be decontaminated. Beyond the conventional techniques for evaluation of the wastes, new emerging techniques and the use of new solvents have drawn attention recently. Among new generation solvents, deep eutectic solvents (DESs) have been increasingly used in the treatment of lignocellulosics to produce value-added products such as biofuels, chemicals, and solvents and also used for the recovery of bioactive phenolic compounds. DESs are used extensively for fractionation of lignocellulosic wastes, often in combination with enzymatic hydrolysis of the biomass. On the other hand, extraction and recovery of bioactive compounds are also under research using DESs. This mini review summarizes the very recent literature reports on the use of DESs in treating agro-industrial wastes within the concept of valorization of biomass.

Keywords: agro-industrial wastes, bioactive phenolic compounds, deep eutectic solvents, lignocellulosic biomass, pretreatment

1. Introduction

Along with the increase of the global consumption manner of the humanity, the general waste amount has been increasing significantly. Global municipal solid waste estimated to increase to 2.2 billion tons annually by the third decade of 2000 [1]. The accumulation of this huge amount of waste creates tedious environmental problems such as the generation of greenhouse gases along with the physical appearance. Despite the studies on the recycling and recovery processes, landfill is still commonly used procedure for the waste disposal in many countries [2].

The main constituent of the municipal solid waste is the lignocellulosic waste having a percentage of 29 [3]. The lignocellulosic waste consists of paper, garden waste, wood, food, and also agricultural wastes. In this chapter, we will focus on the agricultural lignocellulosic waste. A general classification for lignocellulosic waste consists of three subclasses [4], namely, wood leftovers, farming crops, and secondary biomass. Logging leftovers, wastes from pulp, and paper industry are the subclasses of wood leftovers, whereas grasses, short rotation crops, as well as oil and grain crops belong to farming crops. On the other hand, secondary biomass has also two subclasses, namely, municipal solid wastes and food processing wastes.

Lignocellulose represents the matter of plants in general terms. It is the most abundant sustainable carbon source, and the main constituent is lignin that consists of complex organic polymers. Agricultural lignocellulosic biomass is composed of ~35–50% cellulose, 20–35% hemicellulose, and 10–25% lignin [5]. Lignin forms the plant cell walls providing the mechanical endurance to the plant (**Figure 1**). They are mainly composed of monolignols that are methoxylated derivatives of benzene.

The carbohydrates found in the lignin structure are cellulose (**Figure 2**) and hemicellulose (**Figure 3**) that are covalently and hydrogenically bonded to lignin molecules. As a linear-chain polysaccharide, cellulose is made up of D-glucose monomers that are linked with ß-1-4 glycosidic bonds [6]. Hydrogen bonding interactions are present between linear chains that are found in microfibrils [7], and cellulose has several types of crystalline structure. This complex structure provides the rigid and recalcitrance to dissolution of cellulose. Hemicellulose is structurally similar to cellulose as it also consists of polysaccharides, but it has a lower chain amount. On the other hand, hemicellulose contains branched heteropolymer consisting of pentoses – mostly D-xylose and D-arabinose; hexoses – mostly D-mannose, D-glucose, and D-galactose; and sugar acids – mostly 4-*O*-methyl-D-glucuronic acid, D-galacturonic acid, and D-glucuronic acid. Lignocellulosic biomass also contains pectins, proteins, extractives, and ash in low amounts [8]. Lignin has a three-dimensional structure holding the

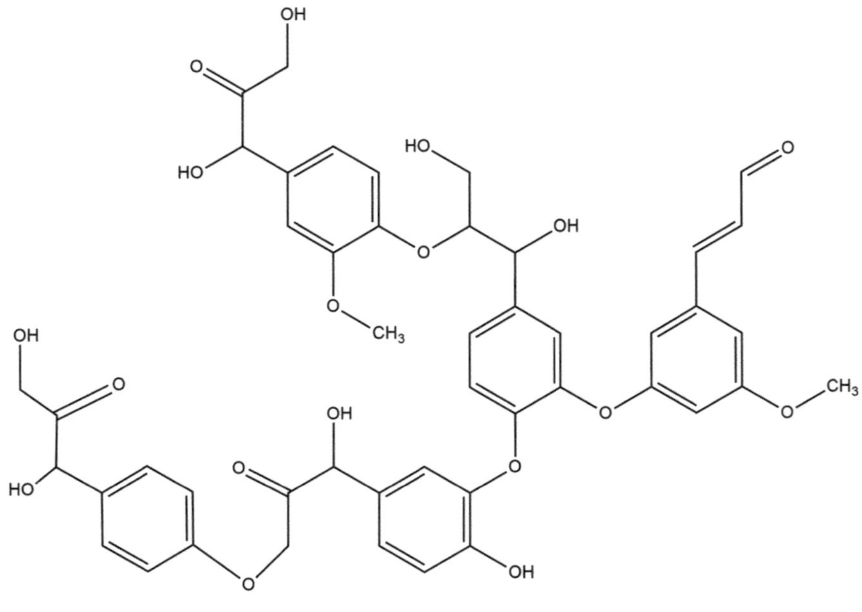

Figure 1.
Chemical structure of lignin.

lignocellulosic structure together, and it is water insoluble [9, 10]. Beyond the massive common information we know about lignin structure, new articles show us [11] that there are still many things to be clarified [12].

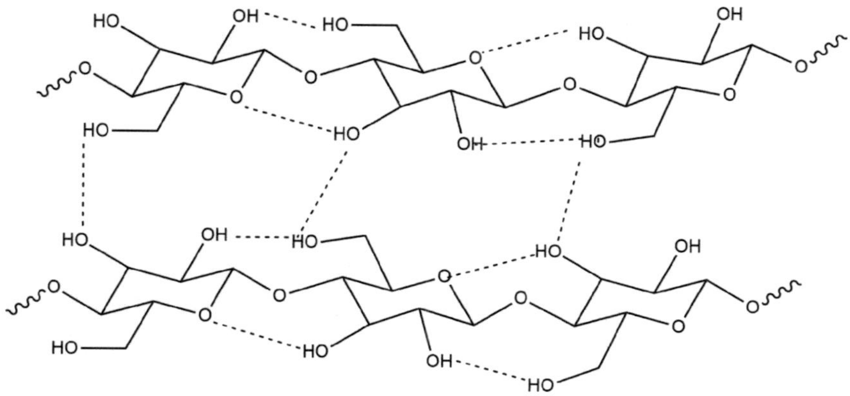

Figure 2.
Chemical structure of cellulose with the schematic illustration of the hydrogen bonding between monomers.

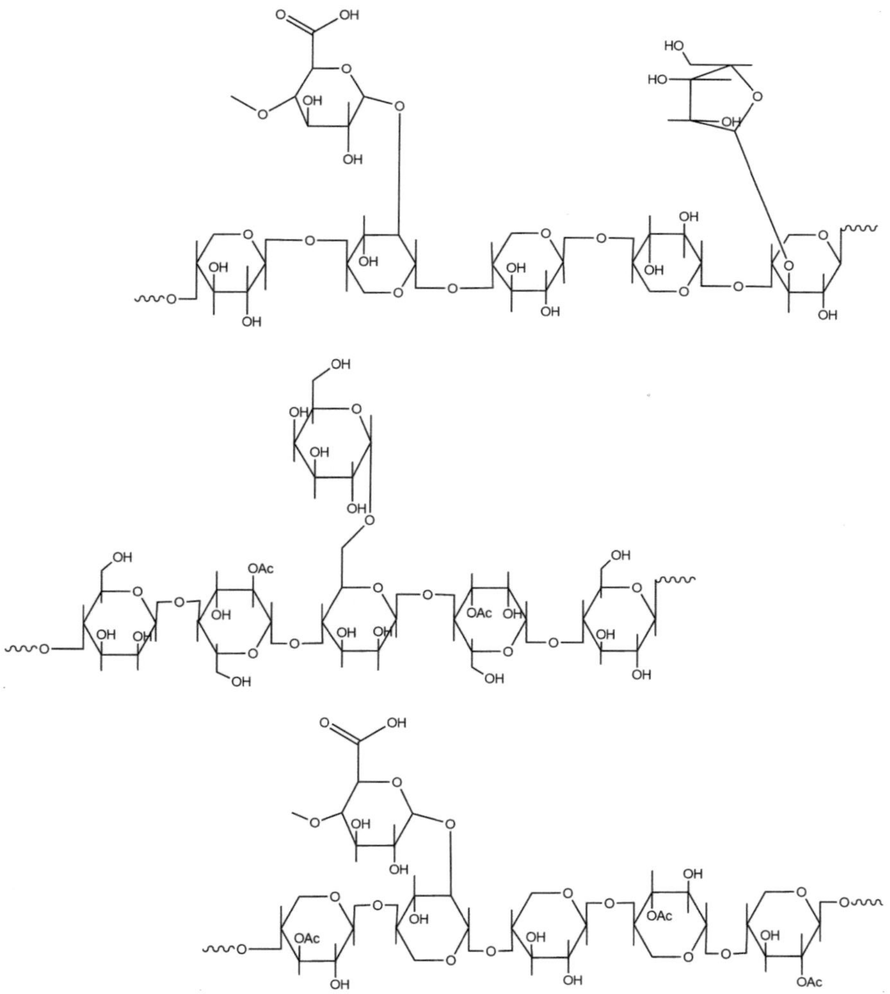

Figure 3.
Three common structures of hemicellulose.

Since lignocellulosic biomass is the most abundant natural source in the world, it may be evaluated as an alternative to unsustainable sources in many aspects. On the other hand, the wastes formed by lignocellulosic materials cause environmental problems arising from organic constituents with high COD and BOD degrees. Most of the lignocellulosic wastes contain phenolic compounds that may cause damage to the environment when discharged without any treatment [13]. These wastes may produce odor, soil pollution, and harborage for insects, if not processed further [14]. A promising approach to reduce the pollution problem of the lignocellulosic biomass is to use them as raw materials as a resource of value-added products such as biofuels (bioethanol, biogas, and biohydrogen), chemicals, and solvents [15] and also to use them for the recovery of bioactive phenolic compounds (flavonoids, phenolic acids, stilbenes, and tannins) [16]. According to the recent literature, the use of lignocellulosic biomass is encouraged as a natural source to be used in biotechnological process that will spontaneously lead to a decrease in the pollution effects of the waste. There are various methods to evaluate the lignocellulosic biomass such as fractionation or recovery of the valuable compounds. Besides conventional methods, the use of green techniques has gained a considerable attention due to environmentally friendly characteristics. In this mini review, the very recent literature on the use of deep eutectic solvents (DESs) in treating agro-industrial wastes, within the concept of valorization of biomass, is summarized.

2. Deep eutectic solvents

For both chemical and pharmaceutical processes, solvents are essential constituents. They are utilized in a broad range of fields including bulk chemicals, medicines, cleaning agents, dyes, and so on. The solvents used in such processes are mainly petroleum-based organic solvents, as well as ammonia and water. However, along with the increasing consciousness related to the environment, the solvents that are regarded as eco-friendly have been the focus for many researchers and are regarded as green solvents. Besides the formerly known green solvents such as bio-solvents and supercritical fluids, ionic liquids and lately deep eutectic solvents have been extensively utilized in various areas with an increasing trend. Additionally, green solvents are encouraged in many fields of research to promote sustainable processes [17].

DESs are one of the most popular green solvents that are mostly known as nontoxic, recyclable, and nonflammable, and they have low vapor pressures [18–20]. They can be easily prepared in the laboratory using numerous substances in different molar ratios that result in diverse properties of DES such as polar, non-polar, acidic, and basic. The most common method to prepare a DES is to mix the constituents in a certain molar ratio at a certain temperature until a homogeneous liquid form is obtained [18, 21] (**Figure 4**). In another method, the constituents (mostly solid) are mixed together with water, and subsequent evaporation of the excess water under vacuum is performed, which is called evaporation method [22]. Similar steps are followed for the freeze-drying method; indeed, water is removed by freeze-drying [23]. In grinding method, a glovebox in nitrogen atmosphere is used to grind the solid in a mortar till clear liquid is obtained [24]. If a novel DES to be formed for the first time, many tests should be performed to prove the "deep eutectic" property of the solvent. Otherwise, published articles' protocols should be followed exactly to synthesize DES in the correct form.

The very first description of DES was made by Abbot et al. as the liquid formed between a variety of quaternary ammonium salts and carboxylic acid. Later on,

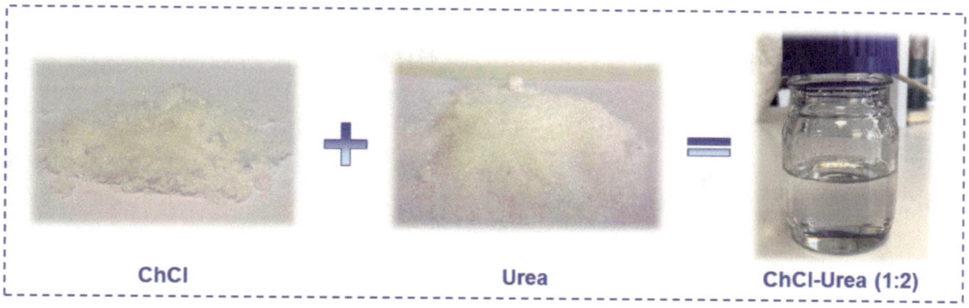

Figure 4.
Choline chloride, urea, and clear liquid DES of ChCl-urea (1:2).

DESs were classified into four groups [25]: Type I: Organic salt + Metal salt; Type II: Organic salt + Metal salt hydrate; Type III: Organic salt + hydrogen bond donor (HBD); and Type IV: Metal salt + HBD. Lately, many researchers presented so many different DESs from so many different types of molecules that the definition of DES converged to a simple form: DESs are composed of two or more components, which in minimum two of them have a hydrogen bonding interaction ability: one as a HBD and one as a hydrogen bond acceptor (HBA) [26]. On the other hand, DESs that are formed by natural compounds such as organic acids, sugars, and choline chloride are called natural deep eutectic solvents (NADESs) [21]. NADESs may be classified as sugar based (glucose-fructose-water, glucose fructose-sucrose-water), polyol based, acid based, and so on. Since DESs can be formed by a number of components, physicochemical properties vary from type to type. Therefore, one can tune the physicochemical property by changing the type and the molar ratio of the constituents. Depending on the type of the constituents, viscosity of DESs may be low or high. High viscosity DESs are hard to be handled, but in some cases, they are preferred to be used as a mixture of alcohol and water to decrease the viscosity. DESs generally have low melting points. This is related to the hydrogen bond interaction between the constituents. Some DESs were reported to have a glass transition temperature [22, 24, 27]. Density ranges of 800–1600 kg/m^3 are presented in the literature, but in general, they have higher density than water [28–30]. On the other hand, hydrophobic DESs are reported to have lower density than hydrophilic DESs [31–33].

The use of DESs in different fields such as biochemistry, electrochemistry, synthesis, nanomaterials, separation, and metal processing [25, 34–38] has been increasing since 2003, when it was first described. Recently, they were shown to be used as solvents in many types of enzyme-catalyzed reactions such as esterification, transesterification, polymerization, and hydrolysis [39–43]. On the other hand, their catalytic effects in several different types of reactions have also been reported [39–41, 44–48]. In detail, the number of DES-related publications was more than 300 between 2009 and 2013, while it was only 29 until 2008 [49]. In 2017, the number of publications on DESs reached up to almost 750 [50].

3. Fractionation of agro-industrial wastes with deep eutectic solvents and recovery of lignin

In the studies carried out within the scope of sustainability, agricultural lignocellulosic wastes such as corn straw, rice straw, wheat straw, fruit wastes, and sunflower stalk have been subjected to various treatments prior to conversion processes. The most challenging step in such a process is the resistance of lignocellulosic material to degradation; therefore, a treatment method is required prior to utilization. These methods can be chemical, physical, mechanical, physicochemical, or biological. In some cases, a combination of these methods is also preferred since each one has different advantages and disadvantages [5]. The most commonly used method is the chemical pretreatment; however, it has undesired environmental impacts. On the other hand, physical pretreatments require high energy, whereas biological pretreatments progress relatively slowly. Therefore, in addition to efficiency, cost, environmental impacts, and ease of use should be taken into consideration for the selection of the pretreatment method.

Ionic liquids as green solvents are effective and promising solvents in the pretreatment of lignocellulosic biomass [31, 51]. However, high prices and toxic properties limit their utilization in industrial applications [31]. Recently, DESs that have superiority to ionic liquids due to their low cost, low volatility, biodegradability, easy preparation techniques, and environmental friendliness have been successfully used in the pretreatment of lignocellulosic materials [50].

Casal et al. [52] were among the first researchers to report that the solubility of wheat stalk in DES was promising. Later on, Francisco et al. [53] studied the solubility of alkali lignin, cellulose, and starch in DESs prepared with choline chloride and carboxylic acid. They reported that lignin was soluble in DESs, whereas cellulose was nearly insoluble, which was a promising result. Among the tested eutectic solvents, the best result was obtained with ChCl-lactic acid (LA) (1:9). Afterward, several researchers treated agro-industrial lignocellulosics with DESs and reported satisfactory results. Procentese et al. [54] pretreated corncob with different choline chloride-based DESs and achieved a total of 41 g fermentable sugars from 100 g corncob after a subsequent enzymatic saccharification. The concentrations of inhibitory agents, that is, acetic acid and furfural were low following the pretreatment with DESs. The authors also reported that the decrease in lignin and hemicellulose contents increased the crystallinity index (CrI) of the pretreated biomass. Zhang et al. [55] pretreated corncob with DESs consisting of choline chloride as HBA and monocarboxylic acid, dicarboxylic acid, or polyalcohol as HBDs. SEM, XRD, and FTIR analyses of treated corncob showed that pretreatment with DESs disrupted the structure of biomass. Polyalcohol-ChCl was found to be more effective in lignin extraction than others. Kumar et al. [56] treated rice straw with lactic acid-betaine and lactic acid-ChCl NADESs and could extract high purity of lignin (>90%). They also reported that approximately 60% of lignin could be separated from the lignocellulosic material. Additionally, higher lignin solubility was achieved when lactic acid-ChCl was used in the treatment. The addition of water (5%) during pretreatment caused a further increase (about 22%) in the extracted amount of lignin. The authors also reported that the CrI of biomass decreased after pretreatment and that subtle structural differences were detected in the crystalline and also amorphous zones of the cellulosic portions. Procentese et al. [57] treated waste lettuce leaves with ChCl-glycerol and used the pretreated biomass sequentially in the enzymatic hydrolysis and acetone-butanol-ethanol fermentation. The authors reported that less energy was consumed with the use of DES than both NaOH and steam explosion pretreatment techniques for the same degree of fragmentation. In the study of rice straw pretreatment using DES, Hou et al. [58] reported that two-step pretreatment increased the yield of sugar by creating a synergism. The researchers found that the yield of glucose was 90.2% as a result of sequential ChCl-oxalic acid and ChCl-urea pretreatments, and also the addition of water during the process increased the yield. Procentese et al. [59] investigated the production of fermentable sugars from biomass by pretreating apple residues, potato peels, coffee silverskin, and brewer's spent grains with ChCl-glycerol and ChCl-ethylene glycol. The highest glucose yield was 0.20 with ChCl-glycerol and 0.19 with ChCl-ethylene glycol. Liu et al. [60] treated wheat straw with triethylbenzyl ammonium chloride/lactic acid (TEBAC/LA)-based deep eutectic solvents under different conditions. The authors reported that the use of TEBAC/LA (1:9) at 373 K for 10 h provided the highest subsequent enzymatic hydrolyses of cellulose and xylan. About 80% removal of lignin was achieved using TEBAC/LA DES in the pretreatment. New et al. [61] investigated the effect of water content of ChCl-urea (1:2) on delignification of oil palm fronds and showed that aqueous DES provided more lignin removal than pure DES. The presence of 30% (v/v) water in DES was reported as the best amount for optimal delignification (16.31%). Ong et al. [62] used two-pot sequential pretreatment for oil palm fronds. They ultrasonicated the palm fronds in water and subsequently pretreated with ChCl-urea. The authors reported that the ultrasound pretreatment facilitated the degradation of lignin matrix by DES. The hydrogen bonding between the halogen component of ChCl and the hydroxyl groups of lignin was proposed to be a facilitation in the cleavage of ether or ester bonds among hemicellulose and lignin. At the optimum conditions (70% amplitude and 30 min), 36.42% of lignin removal and 58% of xylose recovery were achieved. Tan et al. [63] synthesized several DESs using ChCl and organic carboxylic acids and used them in the pretreatment of oil palm empty fruit bunch. It was reported that the presence of hydroxyl moiety and short alkyl chain enhanced the biomass fractionation and lignin extraction. ChCl-LA (1:15) and ChCl-formic acid (1:2) extracted more than 60 wt% of lignin. Fang et al. [64] proposed that a hydrothermal pretreatment could reduce the recalcitrance of lignocellulosic biomass if applied before a DES treatment. The hydrothermal pretreatment was performed at 200°C for 10 min with 10% dry matter loading. The results showed a consistency with the initial proposal. Both xylan and lignin removals were successfully enhanced around 25% during the treatment using ChCl-glycerol (1:2). Similar liquid hot water pretreatment was studied by Tian et al. [65] for the delignification of poplar wood shavings. To provide a mutual agreement for both hemicellulose recovery and solid yield, 170°C was preferred as temperature for the hot water extraction for 40 min. For the subsequent DES treatment step, acidic eutectics were prepared by using ChCl as HBA and formic acid, acetic acid, or lactic acid as HBDs in a molar ratio of 1:2. The hydrothermal processing together with DES treatment increased the lignin selectivity and also the porosity of the resulting cellulose. The ionic properties of the DESs were proposed to provide the selective lignin removal and cellulose deconstruction, thereby increasing cellulose chemical reactivity. A 79.8% of solid yield and 54.4% of hemicellulose removal were reported in the study. Chen et al. [66] aimed to obtain platform chemicals such as furfural, 2,3-butanediol by the pretreatment of switchgrass with ChCl-ethylene glycol. They reported that neat ChCl-ethylene glycol provided a removal of only 24% of lignin, while acidified form provided 87% removal. They also could enrich cellulose up to 72.6% in pretreated switchgrass with the solid loading levels of between 20 and 27%. At this high level of solid-loading efficient, removal of lignin and xylan was achieved. Lim et al. [67] synthesized new DESs using potassium carbonate and glycerol in different molar ratios. The most appropriate molar ratio was reported as 1:7 in terms of pH, viscosity, and thermal stability. They tested different parameters such as temperature (110–150°C), reaction time (40–120 min), and solid-to-liquid ratio (1:8–1:12) on the treatment of rice straw. They could achieve 73.8% cellulose under the optimum conditions that were a temperature of 140°C, a reaction time of 100 min, and a solid-to-liquid mass ratio of 1:10. CrI was reported to increase to 60% from 52.8% after the treatment. Wan and Mun [68] tested the use of different DESs [ChCl-urea (1:2), ChCl-citric acid (1:2), and ChCl-glycerol (1:1)] for the treatment of sago waste. The optimum pretreatment conditions were reported as 110°C and 3 h at 5% solid loading. According to the apparent structural disruption created by

ChCl-urea, it was selected as the DES to give the best result. The authors subsequently performed enzymatic hydrolysis to be mentioned in the next part. A distinct study presented the *in situ* synthesis of DES for the delignification of *Roystonea regia* leaves and leaf sheaths [69]. They claimed that DES could be formed when ChCl was added into water during hydrothermal processing by *in situ* polyhydrogen bonding. The deep eutectic structure was proposed to occur between ChCl as HBA and hemicellulose-derived acids (including formic, acetic, and glucuronic acids) and hydronium ions as HBDs. According to the results, 53.6% lignin removal was obtained for leaf sheaths, while 44.6% was obtained for *Roystonea regia* leaves. They also reported nearly a threefold increase in biomethane yield when *in situ* DES treatment was performed in comparison to hydrothermal processing. Shen et al. [70] reported a reduction in the *Eucalyptus camaldulensis* recalcitrance using ChCl-lactic acid DES. They also declared the preserved structure of important linkages and noncontaminated carbohydrates under the optimum conditions (110°C, 6 h, and 10% solid loading). On the other hand, the increase of lactic acid mole in the DES was reported to cause a relatively lower molecular weight of lignin. In another study, the authors presented the use of diluted alkaline hydrogen peroxide together with DESs in the delignification of oil palm fronds [71]. In the optimum sequential treatment procedure, oil palm fronds were added to 0.25% of alkaline hydrogen peroxide solution at 5% solid loading for 90 min followed by a ChCl-urea (1:2) treatment at 120°C, 4 h, and 10% loading. The authors reported 18.99% of delignification under the optimum conditions with sequential treatment, whereas only 12.16% could be achieved with sole DES treatment. Liu et al. [72] treated moso bamboo using ChCl (1:9) and investigated the effect of temperature (100–120°C), time (2–4 h), and solid-to-liquid mass ratio (1:15–1:25) as parameters in the experimental design. The optimal reaction conditions were determined as a temperature of 120°C, a time of 3 h, and a solid-to-liquid ratio of 1:25. The chemical composition of fibers was reported as 81.4% cellulose, 14.8% hemicellulose, and 3.0% lignin at the optimum conditions. Tan et al. [73] investigated the effect of six different DESs in the single-step fractionation and delignification process of oil palm empty fruit bunch. With this aim, they prepared ChCl-lactic acid (1:5), D-glucose-lactic acid (1:5), ChCl-D-glucose (1:1), ChCl-glycerol (1:2), ChCl-urea (1:2), and potassium carbonate-glycerol (1:6). According to the results, the pH of DESs was reported to have an important effect on the fractionation efficiency. ChCl-lactic acid provided 100% hemicellulose extraction, 88% delignification, and 50% lignin pellet extraction from oil palm empty fruit bunch. Kandanelli et al. [74] presented the utilization of DES together with different cosolvents, such as phloroglucinol, HCl, *n*-butanol, and ethyl acetate, for the efficient delignification of rice husk, rice straw, and wheat straw. They studied the effect of the type of cosolvent, DES-cosolvent ratio (2:1, 1:1, and 1:2), and temperature (50, 80, and 120°C) on the delignification process. The best cosolvent (*n*-butanol) was selected according to the miscibility with DES and also to lignin solubility. The highest delignification was achieved around ~50% using *n*-butanol-assisted DES [ChCl-oxalic acid (1:1) at a ratio of 2:1, with high solid loading of 15% (w/v) at 120°C (~1.2 bar) after 60 min]. They reported about a 2.3-fold increase in the delignification when the temperature increased from 80 to 120°C. A similar usage of DES together with alcohol was also studied by Jablonsky et al. [75] for the delignification of unbleached pulp. Additionally, they synthesized 23 new component DESs and screened for the utilization in the lignin extraction. The authors reported that the addition of alcohol was an advantage in the control of the density and viscosity of DESs. The structure and composition of lignin depended on the pH of the DES used. The best DES for the delignification was reported as malonic acid-ChCl-propanediol (1:1:3), providing 39.80% delignification. On the other hand, all DESs were determined to be selective for cellulose. In another study, fractionation of beech wood polymers was investigated using ChCl-oxalic acid (1:1), ChCl-oxalic acid (1:2), ChCl-potassium hydroxide (1:4), ChCl-lactic acid (1:2), and ChCl-urea (1:2) [76]. The process parameters were investigated in the ranges of 2–24 h time, 60–100°C temperature, and 1:100–1:10 solid-to-liquid mass ratio. The effective DESs on the fractionation were determined as ChCl-oxalic acid (1:1), ChCl-oxalic acid (1:2), and ChCl-potassium hydroxide (1:4). The optimum values for the parameters were reported as 2.5% biomass loading, 6 h time, and 100°C temperature. Muley et al. [77] used microwave-assisted delignification for the treatment of pinewood sawdust. Three different DESs, namely, ChCl-oxalic acid (1:1), ChCl-lactic acid (1:1), and ChCl-formic acid (2:1), were used to test the effect of temperature (110, 130, and 150°C) and time (1, 5, 10, and 15 min) at 2450 MHz. ChCl-oxalic acid and ChCl-formic acid were reported to provide the highest lignin yield. The advantages of the microwave heating were reported as the diminished reaction time in addition to the promotion of selective bond cleavage during lignin depolymerization and a narrow molecular weight distribution. *Eucalyptus globulus* chips were used in the delignification process in which the effect of ChCl in the DES was investigated [78]. Experimental conditions used were a temperature of 120°C, a time of 8 h, and a liquid-to-solid ratio of 20:1. According to the results, chloride anion was determined as the active component of ChCl providing the increase in the cleavage rate of β-O-4 bonds and consequently increasing the delignification rate of biomass. Quek et al. [79] used ChCl-lactic acid, ChCl-glycerol, and ChCl-urea in the ultrasound-assisted pretreatment for the delignification of oil palm empty fruit bunch. The treatment was conducted at 50°C, 10% solid loading, and 240 W for 30 min. The lowest lignin content was obtained using ChCl-lactic acid as 18.8% followed by ChCl-glycerol and ChCl-urea as 19.4 and 21.2%, respectively.

The main idea of the utilization of DES in the pretreatment is the possibility of the strong intramolecular hydrogen bonds in DES to promote the breakage of the hydrogen bonds in the lignocellulosic structure [12, 80–82]. There are also some reports on the mechanism of the interaction between DES and lignocellulosic components. In general aspects, to increase the solubility of a hydrophobic compound in an aqueous solvent, the following well-known methods are utilized, that is, cosolvency, hydrotropy, complexation ionization, and the use of surface-active components [83]. Therefore, the mechanism of the enhanced solubility of lignin in DES-water mixtures is investigated in terms of hydrotropic effect. Such a study was conducted by Soarez et al. [84]. The authors reported that syringic acid solubility was increased by decreasing the polarity of the carboxylic acids in DESs. Apart from the hydrogen bond interactions and pi-pi interactions, the main reason for high solubility was reported as the dispersive interactions between organic acid alkyl chain and syringic acid. Furthermore, when urea was used instead of choline chloride, a fourfold increase in the solubility of lignin was reported. Nearly, 50% of DES-water mixtures provided the best solubility of the monomer. This was explained by the hydrotropic mechanism. They obtained the same result when they used organosolv and kraft lignin and proved the mechanism by dynamic light scattering. Xia et al. [85] searched for the weak fractionation efficiency of ChCl-glycerol using different techniques such as quantum mechanics calculations and solvatochromic parameters. The intramolecular interactions of lignin-carbohydrate complexes were found to be stronger than the interactions with DES and lignin-carbohydrate complexes. Interestingly, chloride ion in DES was reported to be surrounded by mutually anionic hydrogen bonds and cationic hydrogen bonds. This case resulted in a lowered ability of occupied-site anions and insufficient protons, which meant inactive acidic sites. To overcome this, a ternary DES was formed by adding the aluminum chloride into DES. The resulting supramolecular complexes of chlorine ion-metal cation-hydrogen bond acceptor showed deep eutectic characteristics and resulted in a significant enhancement of the efficiency of the lignin extraction. On the other hand, Alvarez-Vasco et al. [86] found that DESs have the ability to cleavage ether bonds without affecting C-C linkages. Considering the ability of the solvents with high ß and π* to provide solubility of lignin, such as dimethylsulfoxide and pyridine, DESs can be possibly declared as new green candidates in lignin solubility.

Aforementioned studies clearly show that the use of DES is a good alternative for the removal of lignin from agro-industrial wastes to conventional pretreatment methods. Subsequent use of these pretreatment products allows the lignocellulosics to be valorized for several industries (**Figure 5**).

3.1 Enzymatic hydrolysis of biomass components

Apart from the treatment studies on the lignocellulosic waste by DESs, additional enzymatic hydrolysis is performed in many studies to remove lignin. Some of the above-mentioned literature contains subsequent enzymatic hydrolysis of the treated biomass as summarized. Procentese et al. [54] performed hydrolysis using Cellic CTec 2 enzymes (Novozyme) after increasing the digestibility of corncob by DES pretreatment. The hydrolysis conditions were 50°C, 180 rpm, and up to 80 h in a rotary shaker. The saccharification rate was found to be the highest at 80°C for the ChCl-imidazole pretreated sample from which 55% lignin was successfully removed. The enzymatic glucose and xylose yields increased with the increasing pretreatment temperature. The highest recovery of the initial carbohydrates was reported as 76%. In their following research, Procentese et al. [57] studied the enzymatic hydrolysis with the same commercial enzyme after the treatment of waste lettuce leaves with DESs. The completion time for the hydrolysis of the pretreated biomass was reported as 9 h. On the other hand, the higher the pretreatment temperature was the higher fraction of monomers was obtained during enzymatic hydrolysis. The authors also used the enzymatic hydrolysate of the pretreated lettuce in the batch culture of *Clostridium acetobutylicum* DSMZ 792 and reported full consumption of the sugars after 60 h. On the other hand, Kumar et al. [56] investigated the saccharification of pretreated rice straw using cellulose. Saccharification efficiency was reported as 36.0 ± 3.2% in 24 h at 10% solid loading. Liu et al. [60] used cellulase and ß-glucosidase for the hydrolysis of DES-pretreated wheat straw. According to the results, 89.06% of cellulose and 71.00% of xylan could be hydrolyzed. Fang et al. [64] proposed a prior treatment using hydrothermal processing for the date palm residues before DES treatment. They reported that the presence of hydrothermal pretreatment provided more efficient hydrolysis using Cellic CTec2. The conversion of glucan to glucose increased by 1.7-fold as a result of the pretreatment. However, cellulose crystallinity was not affected by the pretreatment. Apart from the advantages that hydrothermal processing provided to the process, the

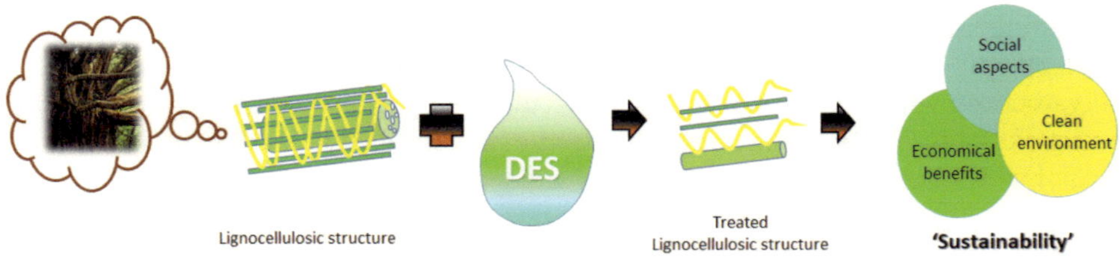

Figure 5.
Schematic illustration of the effect of DES on the lignocellulosic structure to yield sustainable processes.

operating and the investment cost remain a challenge for the process. Chen et al. [66] enzymatically hydrolyzed the pretreated switchgrass with CTec2 and HTec2. They could obtain 206.5 g/L glucose and 34.7 g/L xylose with 86.2% glucose yield within 48 h and reported 90.2 g/L 2,3-butanediol concentration without extra sugar addition for the first time. On the other hand, Wan and Mun [68] performed a subsequent hydrolysis step after the treatment of the sago waste with DESs. The hydrolysis was conducted at 50°C and 100 rpm for 48 h. The highest amount of glucose yield was obtained using ChCl-urea as 5.2 mg/mL. Acidic or alkaline pH DES increased the glucose yield.

4. Extraction and recovery of flavonoids from agro-industrial lignocellulosic wastes

Agro-industrial wastes represent sources of phenolic compounds that have beneficial effects to health due to their antioxidant, antimicrobial, anti-inflammatory, and immune-stimulant properties [87–89]. The prevention of cancer and cardiovascular diseases by phenolic compounds is attributed to their antioxidant and scavenging properties against reactive oxygen species. Apart from their use in biomedical applications, phenolic compounds can also be used in food industry as nutraceuticals. More than 8000 phenolic structures are identified in the structure of plants [90]. The extraction of biophenols from plants is always attractive; moreover, during the last decade, the recovery of phenolic compounds from agro-industrials has gained enormous attention. In spite of their distinct health beneficial properties, the massive phenolic compounds in lignocellulosic wastes have detrimental effects on the environment. The removal/recovery of phenolics from biomass has been conventionally performed with organic solvent extraction; however, recent studies show that DESs can be successfully used in the extraction [91–94]. On the other hand, the use of DESs in biomass processing is less studied than other applications of DESs and needs to be improved [80]. Below, the very recent studies dedicated to the DES selection and condition development for the extraction of polyphenolic bioactive compounds, especially flavonoids, from most abounded agro-industrial wastes are briefly summarized.

Jeong et al. [95] tested several DESs for the recovery of anthocyanin from grape skin and reported that ChCl combined with citric acid, D-(+)-maltose, and fructose was the most effective ChCl-based DESs. In addition, a newly designed DES – citric acid-D-(+)-maltose (4:1) – provided considerably high level extraction yield of anthocyanin. Under the optimized conditions identified by the response surface methodology, total anthocyanin content was found to be 63.36 mg g^{-1} using the new DES. Radosevic et al. [96] used ChCl-based DESs containing glucose, fructose, xylose, glycerol, and malic acid for the recovery of phenolics from grape skin and tested the biological activity of extracts *in vitro* using HeLa and MCF-7 human tumor cell lines. Decreased amount of cytotoxicity of DESs was observed against HeLa and MCF-7 cells. ChCl-malic acid (1:1) provided the highest extraction capability of total phenolic and anthocyanin contents as 91 and 24 mg g^{-1} dw, respectively. Antioxidant activity, in terms of ORAC value, was obtained as 371 mmol TE g^{-1} dw, while antiproliferative activity was around 20%. Recently, Panic et al. [97] reported a scale-up of the extraction process of grape pomace in which anthocyanins were extracted using NADESs. They also studied on the separation of the valuable bioactive compounds from the extracts. ChCl-citric acid (2:1) was successfully used in multimode-microwave and low-frequency-ultrasound irradiation extractions, and it was found that a simultaneous ultrasound/microwave-assisted extraction provided 1.77 mg g^{-1} dw of anthocyanins. Anthocyanins were successfully recovered from NADES and recycled. Chanioti and Tzia [98] used ChCl-citric acid, ChCl-lactic acid, ChCl-maltose, and ChCl-glycerol NADESs to recover bioactive compounds from olive pomace. They added 20% v/v water to the homogenate and tested the effect of different techniques on the extraction process such as high hydrostatic pressure, microwave, and ultrasonication. The NADESs prepared using citric acid and lactic acid were more efficient in the extraction of phenolic compounds of olive pomace than conventional solvent. Ozturk et al. [99] studied the extraction of flavonoids from orange peel using ChCl-glycerol and ChCl-ethylene glycol and reported that ChCl-ethylene glycol (1:4) provided the highest total phenolic compounds (3.61 mg GAE g^{-1}) and also the highest antioxidant activity based on DPPH radical scavenging method (30.6 μg mL^{-1}). The phenolic acids extracted were identified as gallic, *p*-coumaric, ferulic, caffeic, and *trans*-cinnamic acids, as well as flavone and thymol. Ferulic acid was found to be the most abundant phenolic compound, while *p*-coumaric and gallic acids were on the second order. Pal and Jadeja [100] reported the microwave-assisted extraction of polyphenolic compounds from ripe mango (*Mangifera indica* L.) peels with DES consisting of lactic acid-sodium acetate-water (3:1:4). The highest values for the recovery of total phenolic content, ferric reducing antioxidant power, and DPPH scavenging activity were found to be 56.17 mg GAE g^{-1} dw, 683.27 μmol ascorbic acid equivalent g^{-1} dw, and 82.64%, respectively. Mangiferin was detected as the main flavonoid of the extracts. On the other hand, Fernandez et al. [101] designed a new NADES using lactic acid, glucose, and 15% water (LGH-15) and used it in the extraction of phenolic compounds from onion, olive, tomato, and pear industrial byproducts. The results showed that LGH-15 had high extractability characteristic for both high and low polar compounds in comparison to conventional solvents. The stabilizing ability of LGH-15 was found to be quite good since the phenolic compounds could remain stable over 2 months in it. In another study [102], various agro-industrial wastes such as lemon peels, olive leaves, onion solid wastes, red grape pomace, spent filter coffee, and wheat bran were used to test the performance of novel glycerol-based eutectic solvents in the recovery of polyphenolic compounds. Glycerol-ChCl and glycerol-sodium acetate provided high extraction efficiency comparable with aqueous ethanol, while glycerol-sodium-potassium tartrate-water showed lower efficiency in the extraction. The extracts with high polyphenol content also possessed higher reducing power and antiradical activity. Stefou et al. [103] screened sodium propionate-based DESs for the extraction of onion solid wastes, which are rich in quercetin and quercetin conjugates. It was suggested that the use of glycerol-sodium propionate (8:1) could result in high flavonoid content and high antioxidant power in comparison with aqueous glycerol, aqueous ethanol containing citric acid, and aqueous glycerol containing 2-hydroxypropyl β-cyclodextrin. Pal and Jadeja [104] studied the extraction of phenolic antioxidants from onion peel using DESs consisting of ChCl as HBA and sucrose, urea, and sorbitol as HBDs. The optimal conditions for the extraction found by Taguchi's method were reported as a temperature of 60°C, a time of 120 min, and a liquid-to-solid ratio of 50:1 using ChCl-urea (1:2). Under these conditions, total phenolic content was reported as 222.97 mg GAE g^{-1} dw. On the other hand, the extracts obtained using ChCl-sorbitol exhibited comparable DPPH radical scavenging activity to the extracts of aqueous methanol (82.40%). Major flavonoids were identified as quercetin, kaempferol, and myricetin. Very recently, Ruesgas-Ramón et al. [105] used agro-residues from coffee and cocoa industries to recover biomolecules using DESs. The authors reported that the use of lactic acid-ChCl (2:1) provided higher phenolics extraction with ultrasound-probe-assisted extraction than heat stirring-assisted extraction. The main compounds extracted were identified as chlorogenic acid, caffeine, and theobromine. The extraction with DESs from other types of agri-food wastes such as saffron processing wastes [106], olive leaves [102, 107–109], and pigeon pea leaves [110, 111] has also appeared in the recent literature.

The number of studies in open literature on the extraction of bioactive compounds with DESs has been rapidly increasing. Therefore, DESs are easily expected to be used more for the extraction of bioactive phenolic compounds from various sources in the near future. On the other hand, the following issues should be more extensively studied; recovery of phenolics from DESs, stability of phenolics in DESs, reusability of DESs and the scale-up of the extraction processes. Additionally, the determination of biological activities of the phenolic compounds in DES extracts appears to be another field to be focused on in the near feature.

5. Instruments used for the extraction and recovery from lignocellulosic wastes

For the extraction and recovery of lignin or flavonoids from lignocellulosic waste, mostly preferred method is the heating and stirring method [54–61, 63, 65, 76, 78, 99, 103, 105–107, 112] in which oil bath is used to achieve relatively high temperatures. Apart from this conventional method, advanced techniques including the use of ultrasound and microwave irradiations are rarely studied for the extraction and recovery from lignocellulosic wastes. The yield of an extraction from a lignocellulosic material is related to the isolation of the target molecule from the matrix [113]. Ultrasound- and microwave-assisted procedures facilitate the isolation of the target molecules. Microwave heating process was reported to be a very efficient method as it promotes the selective bond cleavage and the stretching of certain bonds at a higher level due to the microwave irradiation [113]. Indeed, in the case of microwave, the disruption of the hydrogen bonds occurs by the dipole rotation of the molecules, and subsequently, dissolved ions migrate that lead to an increase in the penetration of the solvents into the matrix. This procedure results in an easier recovery of the target molecules. Beyond this, the pressure formed by the microwave improves the porosity of the matrix, thereby letting better contact of the solvent with the lignocellulosic material [113]. Related to these positive effects, microwave may be used either in the pretreatment of the biomass or in the extraction process directly. Ultrasound-assisted method is also advantageous when compared to classical methods. Similar to microwave-assisted procedure, it lets a better solvent penetration into the matrix. Ultrasound-assisted extraction creates microsteam in the cells that enhances the mass transfer. Ultrasonication creates longitudinal waves that form alternating compressions in the solvent, which leads to cavitation and gas bubbles. In this period, the bubbles expand, and at a point, the gas in the bubble condenses and a considerable amount of energy is released [114]. With the crashing of the condensed molecules, shock waves occur that lead to very high local temperature and pressure values. Regarding the effectiveness of the two nonconventional methods, we believe that researchers will pay attention to use them in the challenging delignification or extraction processes. Both microwave- and ultrasound-assisted extractions reduce working times and also increase the yield.

6. Different aspects of the utilization of deep eutectic solvents

After the use of DES in both fractionation of lignin and recovery of valuable compounds from lignin, DES can be recycled. The solvents that work as an anti-solvent such as water, ethanol, or acetone result in the solidified form of DES and can be easily removed from the media. After the evaporation or freeze drying of the antisolvent, DES can be obtained in a pure form as mentioned before by several authors [51, 102, 115]. The removal of water in DES can be performed by adding acetone, thereby precipitating DES. The precipitated DES can be reused again by heating.

Beyond the advantages of the use of DES in the treatment of lignocellulosic waste, some drawbacks of DESs were also reported and reviewed in the literature [116]. At high temperatures, DES was reported to decompose to the hydrogen bond donor and acceptor [117]. Moreover, low boiling point component was reported to evaporate. The decomposition temperatures of some DESs were also reported in the literature [117]. Another disadvantage of DESs was reported as the hygroscopicity; however, since certain amount of water is intendedly added during some treatments, this would not be a real disadvantage for many treatments to our opinion.

Another issue is the toxicity of DESs. As green solvents, they are assumed to be nontoxic, especially when natural components are used to form the eutectic mixtures. The most commonly used HBA in DESs is ChCl. Since cholinium is a component of vitamin B complex and the ever first DES was defined using ChCl, ChCl-based DESs are regarded as safe. However, according to the latest research articles, nontoxicity may not be generalized for all types of DESs, including the ones containing the common quaternary ammonium salt. Herein, some of the reports on the toxicity of different types of DESs are summarized.

Hayyan et al. [118] were among the first researchers working on the toxicity of DESs. Four different types of DESs [HBA: ChCl HBDs: glycerine, ethylene glycol, triethylene glycol, and urea (molar ratio: 1:3)] were studied for their toxicity toward *Bacillus subtilis* and *Staphylococcus aureus* as Gram-positive bacteria and *Escherichia coli* and *Pseudomonas aeruginosa* as Gram-negative bacteria. The results showed that DESs were benign for these bacteria. However, the tests for the cytotoxicity of DESs that were performed using *Artemia salina* leach showed higher cytotoxic profile of DESs than their individual components. In their progressive search, Hayyan et al. [119] tested phosphonium-based DESs with different HBDs as glycerine, ethylene glycol, and triethylene glycol (molar ratio: 1:3) for their toxicity and cytotoxicity. They reported that the DESs showed antibacterial activity, and similar to the previous results, DESs and individual components showed different cytotoxicities toward brine shrimp. The authors also searched the toxicity of ammonium-based DESs using glycerine, ethylene glycol, triethylene glycol, and urea as HBDs [120]. The results obtained using *in vitro* cell lines and *in vivo* animal models showed that DESs were more toxic than their individual components; moreover, they enhanced reactive oxygen species production and induced apoptosis in treated cancer cells. However, they were found to be less cytotoxic than ionic liquids. Juneidi et al. [121] prepared different types of DESs [HBA: ChCl, HBDs: ethylene glycol, glycerol, urea (molar ratio, 1:2), and HBA: *N,N* diethyl ethanol ammonium chloride (EAC)-malonic acid (1:1), EAC-zinc nitrate hexahydrate (1:1), EAC-Gly (1:2), EAC-ZnCl2 (1:2)]. They tested the toxicity of DESs toward *Aspergillus niger* in addition to *Cyprinus carpio* fish. According to the results, Type I and II DESs showed higher toxicity than Type III DESs. The authors commented that this was due to the metal salts used in preparing DESs. On the other hand, the toxicity of DESs was found to be related to the concentration of DESs and the nature of the individual components. Similar findings on the toxicity of ChCl-ZnCl2 and acidic type of DESs were also reported by Juneidi et al. [122] in their progressive research. The largest inhibition zone for the tested fungi (*Phanerochaete chrysosporium, Aspergillus niger, Lentinus tigrinus, and Candida cylindracea*) was obtained by ChCl-ZnCl2 (1:2), followed by the acidic DESs, ChCl-*p*-toluene sulfonic acid (1:3), and ChCl-malonic acid (1:1). Hayyan et al. [123] claimed that ChCl-malonic acid

(1:1) was more toxic than the other acidic DESs they tested [ChCl-fructose-water (5:2:5), ChCl-glucose-water (5:2:5), ChCl-sucrose-water (4:1:4), and ChCl-glycerol-water (1:2:1)]. Glycerol, fructose, glucose, and sucrose containing NADESs showed significantly low toxicity as expected due to their nature, whereas the organic acid containing DES was quite toxic. Additionally, they reported that a trend could be observed between the cytotoxicity profile and the viscosity, the water content, and the nature of the components of NADESs. Ahmadi et al. [124] presented a research on the assessment of the cytotoxicity of 28 NADESs (HBA: ChCl, HBDs: ethylene glycol, glycerol, 1,2-propanediol, sorbitol, xylitol, xylose, fructose, glucose, mannose, glucosamine, sucrose, maltose, rhamnose, raffinose) by using quantitative structure activity relationship analysis. According to their findings, an increase in the glycerol molar ratio decreased the toxic effect of NADES, whereas an increase in the 1,2-propanediol molar ratio increased the toxic effect. On the other hand, the common ionic liquid [C8mim] [Cl] was found to be more toxic than all NADESs tested. This result was consistent with other studies compared the toxicity of different ILs with DESs [125, 126]. Macario et al. [127] prepared 23 DESs using ethylene glycol and 1-propanol as HBDs; [N1111]Cl, [N2222]Cl, and [N3333]Cl as HBAs, at four different molar ratios 1:2, 1:1, 2:1, and 4:1. They reported that these DESs were not harmful to *Aliivibrio fischeri*, whereas they showed the toxicity in the following order: [N1111]Cl-based DES < [N2222]Cl-based DES < [N3333]Cl-based DES. Mbous et al. [128] compared the cytotoxicity of NADESs prepared using choline chloride, fructose, or glucose with a DES (*N*,*N*-diethyl ethanolammonium chloride:triethylene glycol). NADESs were found to be less toxic than DES. Radosevic et al. [129] prepared 10 DESs as ChCl-oxalic acid (1:1), ChCl-urea (1:2), ChCl-xylitol (5:2), ChCl-sorbitol (2:3), betaine-glucose (5:2), betaine-malic acid-proline (1:1:1), betaine-malic acid-glucose (1:1:1), citric acid-proline (1:1), citric acid-glucose-glycerol (1:1:1), citric acid-fructose-glycerol (1:1:1). Citric acid-based NADESs did not show any cytotoxic effects on the tested cells, whereas ChCl-urea was toxic to only one kind of the cells (MCF-7 cells).

Regarding the above-mentioned reports, it can be concluded that the toxicity of a DES varies according to the type and molar ration of the components. The presence of an organic acid increases the overall toxicity of a DES. The eutectic mixture is generally more toxic than its individual components. This may be related to hydrogen bonds between the components. On the other hand, NADESs were shown to exhibit lower toxicity than DESs; moreover, natural based or not, all DESs exhibit lower toxicity than ILs. Therefore, "green solvent" characteristic should be proved, or the proven green solvents should be used especially in large-scale applications. On the other hand, new strategies should be developed to predict the toxicity level of a candidate green solvent in the near future.

7. Conclusions

Recent trends in the reduction of pollution effects of lignocellulosic waste are to convert them into useful products or to recover the natural components within them. Treatment of wastes provides the utilization of materials or substrates in the production of value-added products. However, most of the pretreatment methods used for lignocellulosics are energy consuming operations since they need to use high temperature and pressure for the removal of lignin. Deep eutectic solvents are new generation solvents that find several applications in chemistry and chemical engineering. Extraction methodology for the recovery of bioactive phenolic compounds from plants can also be used for agro-industrial wastes. Treatment with DESs requires less energy. Additionally, DESs facilitate accessibility to cellulose by dissolving lignin at low temperatures and pressures. On the other hand, the task-specific DES for the target lignocellulosic waste differs according to the type of the waste. Therefore, one should search the literature and start with the promising DES and extend the research to similar kind of DESs to achieve high recovery yields. Apart from that, a new and target-specific DES may be synthesized since enormous number and type of components are candidates for new DESs.

DESs have unique properties such as polarity, conductivity, and viscosity depending on their composition. Therefore, novel DESs to be prepared and treatment conditions to be improved will help to solve environmental problems originated from agro-industrial wastes and also to develop new platforms for the production of valuable products such as chemicals, biofuels, and bioactive phenolic compounds.

Even promising results are published on the use of DESs for the lignocellulosic wastes, the mechanism of the treatment and the changes on the structure of DES still need to be clarified. Additionally, a detailed structural analysis on the extracted and purified biomass components relevant for the purpose should be revisited.

Conflict of interest

The authors declare no conflict of interest.

Author details

Ayşe Ezgi Ünlü and Serpil Takaç*
Faculty of Engineering, Department of Chemical Engineering, Ankara University, Tandoğan, Ankara, Turkey

*Address all correspondence to: takac@eng.ankara.edu.tr

References

[1] Hoornweg D, Bhada-Tat P. What a waste: A global review of solid waste management. Urban Development Series Knowledge Papers. 2012;**15**:1-98

[2] Zheng W, Phoungthong K, Lü F, Shao LM, He PJ. Evaluation of a classification method for biodegradable solid wastes using anaerobic degradation parameters. Waste Management. 2013;**33**:2632-2640

[3] EPA. Municipal Solid Waste (MSW) in the United States: Facts and Figures. Washington, D.C.; 2010

[4] Tan KT, Lee KT, Mohamed AR. Role of energy policy in renewable energy accomplishment: The case of second-generation bioethanol. Energy Policy. 2008;**36**:3360-3365

[5] Isikgor FH, Becer CR. Lignocellulosic biomass: A sustainable platform for the production of bio-based chemicals and polymers. Polymer Chemistry. 2015;**6**:4497-4559

[6] Brandt A, Gräsvik J, Hallett JP, Welton T. Deconstruction of lignocellulosic biomass with ionic liquids. Green Chemistry. 2013;**15**(3):537-848

[7] Sun N, Rodríguez H, Rahman M, Rogers RD. Where are ionic liquid strategies most suited in the pursuit of chemicals and energy from lignocellulosic biomass? Chemical Communications. 2011;**47**:1405-1421

[8] Kumar P, Barrett DM, Delwiche MJ, Stroeve P. Methods for pretreatment of lignocellulosic biomass for efficient hydrolysis and biofuel production. Industrial and Engineering Chemistry Research. 2009;**48**:3713-3729

[9] Zakzeski J, Bruijnincx PCA, Jongerius AL, Weckhuysen BM. The catalytic valorization of lignin for the production of renewable chemicals. Chemical Reviews. 2010;**110**(6):3552-3599

[10] Adorjan I, Potthast A, Rosenau T, Sixta H, Kosma P. Discoloration of cellulose solutions in N-methylmorpholine-N-oxide (lyocell). Part 1: Studies on model compounds and pulps. Cellulose. 2005;**12**(1):51-57

[11] Ralph J, Lapierre C, Boerjan W. Lignin structure and its engineering. Current Opinion in Biotechnology. 2019;**56**:240-249

[12] Van Osch DJGP, Kollau LJBM, Van Den Bruinhorst A, Asikainen S, Rocha MAA, Kroon MC. Ionic liquids and deep eutectic solvents for lignocellulosic biomass fractionation. Physical Chemistry Chemical Physics. 2017;**19**:2636-2665

[13] Niaounakis M, Halvadakis CP. Olive processing waste management. In: Literature Review and Patent Survey. 2nd ed. Pergamon: Elsevier; 2006

[14] Shalini R, Gupta DK. Utilization of pomace from apple processing industries: A review. Journal of Food Science and Technology. 2010;47(4):365-371

[15] Tuck CO, Pérez E, Horváth IT, Sheldon RA, Poliakoff M. Valorization of biomass: Deriving more value from waste. Science. 2012;**337**:695-699

[16] Li X, Row KH. Development of deep eutectic solvents applied in extraction and separation. Journal of Separation Science. 2016;**39**(18):3505-3520

[17] Li Y, Fabiano-Tixier AS, Tomao V, Cravotto G, Chemat F. Green ultrasound-assisted extraction of carotenoids based on the bio-refinery concept using sunflower oil as an alternative solvent. Ultrasonics Sonochemistry. 2013;**20**(1):12-18

[18] Abbott AP, Capper G, Davies DL, Rasheed RK, Tambyrajah V. Novel solvent properties of choline chloride/urea mixtures. Chemical Communications. 2003;**1**:70-71

[19] Zhang Q, De Oliveira VK, Royer S, Jerome F. Deep eutectic solvents: Syntheses, properties and applications. Chemical Society Reviews. 2012;**41**:7109-7146

[20] García G, Aparicio S, Ullah R, Atilhan M. Deep eutectic solvents: Physicochemical properties and gas separation applications. Energy & Fuels. 2015;**29**(4):2616-2644

[21] Dai Y, Witkamp GJ, Verpoorte R, Choi YH. Natural deep eutectic solvents as a new extraction media for phenolic metabolites in carthamus tinctorius L. Analytical Chemistry. 2013;**85**(13):6272-6278

[22] Dai Y, van Spronsen J, Witkamp GJ, Verpoorte R, Choi YH. Natural deep eutectic solvents as new potential media for green technology. Analytica Chimica Acta. 2013;**766**:61

[23] Gutiérrez MC, Ferrer ML, Mateo CR, Del Monte F. Freeze-drying of aqueous solutions of deep eutectic solvents: A suitable approach to deep eutectic suspensions of self-assembled structures. Langmuir. 2009;**25**(10):5509-5515

[24] Florindo C, Oliveira FS, Rebelo LPN, Fernandes AM, Marrucho IM. Insights into the synthesis and properties of deep eutectic solvents based on cholinium chloride and carboxylic acids. ACS Sustainable Chemistry & Engineering. 2014;**2**(10):2416-2425

[25] Smith EL, Abbott AP, Ryder KS. Deep eutectic solvents (DESs) and their applications. Chemical Reviews. 2014;**114**(21):11060-11082

[26] Durand E, Lecomte J, Villeneuve P. Deep eutectic solvents: Synthesis, application, and focus on lipase-catalyzed reactions. European Journal of Lipid Science and Technology. 2013;**115**(4):379-385

[27] Zubeir LF, Lacroix MHM, Kroon MC. Low transition temperature mixtures as innovative and sustainable CO_2 capture solvents. The Journal of Physical Chemistry. B. 2014;**118**(49):14429-14441

[28] Dean PM, Pringle JM, MacFarlane DR. Structural analysis of low melting organic salts: Perspectives on ionic liquids. Physical Chemistry Chemical Physics. 2010;**12**:9144-9153

[29] Fredlake CP, Crosthwaite JM, Hert DG, Aki SNVK, Brennecke JF. Thermophysical properties of imidazolium-based ionic liquids. Journal of Chemical & Engineering Data. 2004;**49**(4):954-964

[30] Almeida HFD, Passos H, Lopes-Da-Silva JA, Fernandes AM, Freire MG, Coutinho JAP. Thermophysical properties of five acetate-based ionic liquids. Journal of Chemical & Engineering Data. 2012;**57**(11):3005-3013

[31] Van Osch DJGP, Zubeir LF, Van Den Bruinhorst A, Rocha MAA, Kroon MC. Hydrophobic deep eutectic solvents as water-immiscible extractants. Green Chemistry. 2015;**17**:4518-4521

[32] Van Osch DJGP, Parmentier D, Dietz CHJT, Van Den Bruinhorst A, Tuinier R, Kroon MC. Removal of alkali and transition metal ions from water with hydrophobic deep eutectic solvents. Chemical Communications. 2016;**52**:11987-11990

[33] Ribeiro BD, Florindo C, Iff LC, Coelho MAZ, Marrucho IM. Menthol-based eutectic mixtures: Hydrophobic low viscosity solvents. ACS Sustainable Chemistry & Engineering. 2015;**3**(10):2469-2477

[34] Zainal-Abidin MH, Hayyan M, Hayyan A, Jayakumar NS. New horizons in the extraction of bioactive compounds using deep eutectic solvents: A review. Analytica Chimica Acta. 2017;**979**:1-23. DOI: 10.1016/j.aca.2017.05.012

[35] Shishov A, Bulatov A, Locatelli M, Carradori S, Andruch V. Application of deep eutectic solvents in analytical chemistry. A review. Microchemical Journal. 2017;**135**:33-38

[36] Hadj-Kali M. Separation of ethylbenzene and n-octane using deep eutectic solvents. Green Processing and Synthesis. 2015;**4**(2):117-123

[37] Lee JS. Deep eutectic solvents as versatile media for the synthesis of noble metal nanomaterials. Nanotechnology Reviews. 2017;**6**(3):271-278

[38] Isaifan RJ, Amhamed A. Review on carbon dioxide absorption by choline chloride/urea deep eutectic solvents. Advances in Chemotherapy. 2018;**2018**:1-6

[39] Wang L, Zhou M, Chen Q, He M-Y. Brønsted acidic deep eutectic solvent catalysed the one-pot synthesis of 2H-indazolo[2,1-b]phthalazine-triones. Journal of Chemical Research. 2013;**37**(10):598-600

[40] Tran PH, Nguyen HT, Hansen PE, Le TN. An efficient and green method for regio-and chemo-selective Friedel–Crafts acylations using a deep eutectic solvent ([CholineCl][$ZnCl_2$]$_3$). RSC Advances. 2016;**6**(43):37031-37038

[41] De Santi V, Cardellini F, Brinchi L, Germani R. Novel Brønsted acidic deep eutectic solvent as reaction media for esterification of carboxylic acid with alcohols. Tetrahedron Letters. 2012;**53**(38):5151-5155

[42] Ünlü AE, Prasad B, Anavekar K, Bubenheim P, Liese A. Investigation of a green process for the polymerization of catechin. Preparative Biochemistry & Biotechnology. 2017;**47**(9):918-924

[43] Arıkaya A, Ünlü AE, Takaç S. Use of deep eutectic solvents in the enzyme catalysed production of ethyl lactate. Process Biochemistry. 2019;**84**:53-59

[44] Musale RM, Shukla SR. Deep eutectic solvent as effective catalyst for aminolysis of polyethylene terephthalate (PET) waste. International Journal of Plastics Technology. 2016;**20**(1):106-120

[45] Patil UB, Singh AS, Nagarkar JM. Choline chloride based eutectic solvent: An efficient and reusable solvent system for the synthesis of primary amides from aldehydes and from nitriles. RSC Advances. 2014;**4**(3):1102-1106

[46] Mondal D, Sharma M, Wang C-H, Lin Y-C, Huang H-C, Saha A, et al. Deep eutectic solvent promoted one step sustainable conversion of fresh seaweed biomass to functionalized graphene as a potential electrocatalyst. Green Chemistry. 2016;**18**(9):2819-2826

[47] Singh B, Lobo H, Shankarling G. Selective N-alkylation of aromatic primary amines catalyzed by bio-catalyst or deep eutectic solvent. Catalysis Letters. 2011;**141**(1):178-182

[48] Keshavarzipour F, Tavakol H. Deep eutectic solvent as a recyclable catalyst for three-component synthesis of β-amino carbonyls. Catalysis Letters. 2015;**145**(4):1062-1066

[49] Tang B, Row KH. Recent developments in deep eutectic solvents in chemical sciences. Monatshefte für Chemie. 2013;**144**(10):1427-1454

[50] Satlewal A, Agrawal R, Bhagia S, Sangoro J, Ragauskas AJ. Natural deep eutectic solvents for lignocellulosic biomass pretreatment: Recent developments, challenges and novel

opportunities. Biotechnology Advances. 2018;**36**(8):2032-2050

[51] Luo J, Cai M, Gu T. Pretreatment of lignocellulosic biomass using green ionic liquids. In: Gu T, editor. Green Biomass Pretreatment for Biofuels Production. Dordrecht: Springer, Dordrecht; 2013

[52] Casal F, Bruinhorts A, Rocha MA, Kroon MC. Suitability of deep eutectic solvents (organic salt and natural carboxylic acid mixtures) for lignocellulosic biomass processing. In: 26th European Symposium on Applied Thermodynamics. Potsdam; 2012

[53] Francisco M, van den Bruinhorst A, Kroon MC. New natural and renewable low transition temperature mixtures (LTTMs): Screening as solvents for lignocellulosic biomass processing. Green Chemistry. 2012;**14**(8):2153-2157

[54] Procentese A, Johnson E, Orr V, Garruto Campanile A, Wood JA, Marzocchella A, et al. Deep eutectic solvent pretreatment and subsequent saccharification of corncob. Bioresource Technology. 2015;**192**:31-36

[55] Zhang CW, Xia SQ, Ma PS. Facile pretreatment of lignocellulosic biomass using deep eutectic solvents. Bioresource Technology. 2016;**219**:1-5

[56] Kumar AK, Parikh BS, Pravakar M. Natural deep eutectic solvent mediated pretreatment of rice straw: Bioanalytical characterization of lignin extract and enzymatic hydrolysis of pretreated biomass residue. Environmental Science and Pollution Research. 2016;**23**(10):9265-9275

[57] Procentese A, Raganati F, Olivieri G, Russo ME, Rehmann L, Marzocchella A. Low-energy biomass pretreatment with deep eutectic solvents for bio-butanol production. Bioresource Technology. 2017;**243**:464-473

[58] Hou XD, Feng GJ, Ye M, Huang CM, Zhang Y. Significantly enhanced enzymatic hydrolysis of rice straw via a high-performance two-stage deep eutectic solvents synergistic pretreatment. Bioresource Technology. 2017;**238**:139-146

[59] Procentese A, Raganati F, Olivieri G, Russo ME, Rehmann L, Marzocchella A. Deep eutectic solvents pretreatment of agro-industrial food waste. Biotechnology for Biofuels. 2018;**11**(37):1-12

[60] Liu Y, Zheng J, Xiao J, He X, Zhang K, Yuan S, et al. Enhanced enzymatic hydrolysis and lignin extraction of wheat straw by triethylbenzyl ammonium chloride/ lactic acid-based deep eutectic solvent pretreatment. ACS Omega. 2019;**4**:19829-19839

[61] New EK, Wu TY, Tien Loong Lee CB, Poon ZY, Loow YL, Wei Foo LY, et al. Potential use of pure and diluted choline chloride-based deep eutectic solvent in delignification of oil palm fronds. Process Safety and Environment Protection. 2019;**123**:190-198

[62] Ong VZ, Wu TY, Lee CBTL, Cheong NWR, Shak KPY. Sequential ultrasonication and deep eutectic solvent pretreatment to remove lignin and recover xylose from oil palm fronds. Ultrasonics Sonochemistry. 2019;**58**:1-10

[63] Tan YT, Ngoh GC, Chua ASM. Effect of functional groups in acid constituent of deep eutectic solvent for extraction of reactive lignin. Bioresource Technology. 2019;**281**:359-366

[64] Fang C, Thomsen MH, Frankær CG, Brudecki GP, Schmidt JE, Alnashef IM. Reviving pretreatment effectiveness of deep eutectic solvents on lignocellulosic date palm residues by prior recalcitrance reduction. Industrial

and Engineering Chemistry Research. 2017;**56**:3167-3174

[65] Tian D, Guo Y, Hu J, Yang G, Zhang J, Luo L, et al. Acidic deep eutectic solvents pretreatment for selective lignocellulosic biomass fractionation with enhanced cellulose reactivity. International Journal of Biological Macromolecules. 2020;**142**:288-297

[66] Chen Z, Bai X, Lusi A, Wan C. High-solid lignocellulose processing enabled by natural deep eutectic solvent for lignin extraction and industrially relevant production of renewable chemicals. ACS Sustainable Chemistry & Engineering. 2018;**6**:12205-12216

[67] Lim WL, Gunny AAN, Kasim FH, AlNashef IM, Arbain D. Alkaline deep eutectic solvent: A novel green solvent for lignocellulose pulping. Cellulose. 2019;**26**:4085-4098

[68] Wan YL, Mun YJ. Assessment of natural deep eutectic solvent pretreatment on sugar production from lignocellulosic biomass. In: MATEC Web of Conferences. 2018. pp. 1-10

[69] Yu Q, Qin L, Liu Y, Sun Y, Xu H, Wang Z, et al. In situ deep eutectic solvent pretreatment to improve lignin removal from garden wastes and enhance production of bio-methane and microbial lipids. Bioresource Technology. 2019;**271**:210-217

[70] Shen XJ, Wen JL, Mei QQ, Chen X, Sun D, Yuan TQ, et al. Facile fractionation of lignocelluloses by biomass-derived deep eutectic solvent (DES) pretreatment for cellulose enzymatic hydrolysis and lignin valorization. Green Chemistry. 2019;**21**:275-283

[71] Ho MC, Wu TY, Chee SWQ, Ngang CY, Chew IML, Teoh WH, et al. An application of low concentration alkaline hydrogen peroxide at non-severe pretreatment conditions together with deep eutectic solvent to improve delignification of oil palm fronds. Cellulose. 2019;**26**:8557-8573

[72] Liu Q, Yuan T, Jin FQ, Yuan BY, Peng F, Li YC. Choline chloride-lactic acid deep eutectic solvent for delignification and nanocellulose production of moso bamboo. Cellulose. 2019;**26**:9447-9462

[73] Tan YT, Ngoh GC, Chua ASM. Evaluation of fractionation and delignification efficiencies of deep eutectic solvents on oil palm empty fruit bunch. Industrial Crops and Products. 2018;**123**:271-277

[74] Kandanelli R, Thulluri C, Mangala R, Rao PVC, Gandham S, Velankar HR. A novel ternary combination of deep eutectic solvent-alcohol (DES-OL) system for synergistic and efficient delignification of biomass. Bioresource Technology. 2018;**265**:573-576

[75] Jablonsky M, Majova V, Ondrigova K, Sima J. Preparation and characterization of physicochemical properties and application of novel ternary deep eutectic solvents. In: Cellulose. 2019. pp. 3031-3045

[76] Mamilla JLK, Novak U, Grilc M, Likozar B. Natural deep eutectic solvents (DES) for fractionation of waste lignocellulosic biomass and its cascade conversion to value-added bio-based chemicals. Biomass and Bioenergy. 2019;**120**:417-425

[77] Muley PD, Mobley JK, Tong X, Novak B, Stevens J, Moldovan D, et al. Rapid microwave-assisted biomass delignification and lignin depolymerization in deep eutectic solvents. Energy Conversion and Management. 2019;**196**:1080-1088

[78] Smink D, Juan A, Schuur B, Kersten SRA. Understanding the role of choline chloride in deep eutectic solvents used for biomass delignification. Industrial and Engineering Chemistry Research. 2019;**58**(66):16348-16357

[79] Quek JD, Lee KM, Lim S, Tey WY, Kang HS, Quen LK. Delignification of oil palm empty fruit bunch via ultrasound-assisted deep eutectic solvent pretreatment. In: International Conference on Sustainable Energy and Green Technology. IOP Publishing; 2019, 2020. pp. 1-6

[80] Loow YL, New EK, Yang GH, Ang LY, Foo LYW, Wu TY. Potential use of deep eutectic solvents to facilitate lignocellulosic biomass utilization and conversion. Cellulose. 2017;**24**:3591-3618

[81] Tang X, Zuo M, Li Z, Liu H, Xiong C, Zeng X, et al. Green processing of lignocellulosic biomass and its derivatives in deep eutectic solvents. ChemSusChem. 2017;**10**(13):2696-2706

[82] Zdanowicz M, Wilpiszewska K, Spychaj T. Deep eutectic solvents for polysaccharides processing. A review. Carbohydrate Polymers. 2018;**15**:361-380

[83] Jouyban A. Handbook of Solubility Data for Pharmaceuticals. Boca Raton: CRC Press; 2009

[84] Soares B, Tavares DJP, Amaral JL, Silvestre AJD, Freire CSR, Coutinho JAP. Enhanced solubility of lignin monomeric model compounds and technical lignins in aqueous solutions of deep eutectic solvents. ACS Sustainable Chemistry & Engineering. 2017;**5**:4056-4065

[85] Xia Q, Liu Y, Meng J, Cheng W, Chen W, Liu S, et al. Multiple hydrogen bond coordination in three-constituent deep eutectic solvents enhances lignin fractionation from biomass. Green Chemistry. 2018;**20**:2711-2721

[86] Alvarez-Vasco C, Ma R, Quintero M, Guo M, Geleynse S, Ramasamy KK, et al. Unique low-molecular-weight lignin with high purity extracted from wood by deep eutectic solvents (DES): A source of lignin for valorization. Green Chemistry. 2016;**18**:5133-5141

[87] Balasundram N, Sundram K, Samman S. Phenolic compounds in plants and agri-industrial by-products: Antioxidant activity, occurrence, and potential uses. Food Chemistry. 2006;**99**(1):191-203

[88] Vodnar DC, Călinoiu LF, Dulf FV, Ştefănescu BE, Crişan G, Socaciu C. Identification of the bioactive compounds and antioxidant, antimutagenic and antimicrobial activities of thermally processed agro-industrial waste. Food Chemistry. 2017;**231**:131-140

[89] El-hawary SS, Rabeh MA. *Mangifera indica* peels: A common waste product with impressive immunostimulant , anticancer and antimicrobial potency. Journal of Natural Science Research. 2014;**4**(3):102-115

[90] Tanase C, Cosarcă S, Muntean DL. A critical review of phenolic compounds extracted from the bark of woody vascular plants and their potential biological activity. Molecules. 2019;**24**(6):1182

[91] Ruesgas-Ramón M, Figueroa-Espinoza MC, Durand E. Application of deep eutectic solvents (DES) for phenolic compounds extraction: Overview, challenges, and opportunities. Journal of Agricultural and Food Chemistry. 2017;**65**(18):3591-3601

[92] Mišan A, Nađpal J, Stupar A, Pojić M, Mandić A, Verpoorte R, et al. The perspectives of natural deep

eutectic solvents in agri-food sector. Critical Reviews in Food Science and Nutrition. 2019;**59**:1-29. DOI: 10.1080/10408398.2019.1650717

[93] Torres-Valenzuela LS, Ballesteros-Gomez A, Rubio S. Green solvents for the extraction of high added-value compounds form agri-food waste. Food Engineering Reviews. 2020;**12**:83-100. DOI: 10.1007/s12393-019-09206-y

[94] Kalhor P, Ghandi K. Deep eutectic solvents for pretreatment, extraction, and catalysis of biomass and food waste. Molecules. 2019;**24**(22):4012

[95] Jeong KM, Zhao J, Jin Y, Heo SR, Han SY, Yoo DE, et al. Highly efficient extraction of anthocyanins from grape skin using deep eutectic solvents as green and tunable media. Archives of Pharmacal Research. 2015;**38**(12):2143-2152

[96] Radošević K, Ćurko N, Gaurina Srček V, Cvjetko Bubalo M, Tomašević M, Kovačević Ganić K, et al. Natural deep eutectic solvents as beneficial extractants for enhancement of plant extracts bioactivity. LWT- Food Science and Technology. 2016;**73**:45-51

[97] Panić M, Radić Stojković M, Kraljić K, Škevin D, Radojčić Redovniković I, Gaurina Srček V, et al. Ready-to-use green polyphenolic extracts from food by-products. Food Chemistry. 2019;**283**(December 2018):628-636. DOI: 10.1016/j.foodchem.2019.01.061

[98] Chanioti S, Tzia C. Extraction of phenolic compounds from olive pomace by using natural deep eutectic solvents and innovative extraction techniques. Innovative Food Science and Emerging Technologies. 2018;**48**(June):228-239

[99] Ozturk B, Parkinson C, Gonzalez-Miquel M. Extraction of polyphenolic antioxidants from orange peel waste using deep eutectic solvents. Separation and Purification Technology. 2018;**206**:1-13

[100] Pal CBT, Jadeja GC. Microwave-assisted extraction for recovery of polyphenolic antioxidants from ripe mango (*Mangifera indica* L.) peel using lactic acid/sodium acetate deep eutectic mixtures. Food Science and Technology International. 2019;**26**(1):78-92

[101] Fernández d l ÁM, Espino M, FJV G, Silva MF. Novel approaches mediated by tailor-made green solvents for the extraction of phenolic compounds from agro-food industrial by-products. Food Chemistry. 2018;**239**:671-678

[102] Mouratoglou E, Malliou V, Makris DP. Novel glycerol-based natural eutectic mixtures and their efficiency in the ultrasound-assisted extraction of antioxidant polyphenols from agri-food waste biomass. Waste and Biomass Valorization. 2016;**7**(6):1377-1387

[103] Stefou I, Grigorakis S, Loupassaki S, Makris DP. Development of sodium propionate-based deep eutectic solvents for polyphenol extraction from onion solid wastes. Clean Technologies and Environmental Policy. 2019;**21**(8):1563-1574

[104] Pal CBT, Jadeja GC. Deep eutectic solvent-based extraction of polyphenolic antioxidants from onion (*Allium cepa* L.) peel. Journal of the Science of Food and Agriculture. 2019;**99**(4):1969-1979

[105] Ruesgas-Ramón M, Suárez-Quiroz ML, González-Ríos O, Baréa B, Cazals G, Figueroa-Espinoza MC, et al. Biomolecules extraction from coffee and cocoa by- and co-products using deep eutectic solvents. Journal of the Science of Food and Agriculture. 2020;**100**:81-91

[106] Lakka A, Grigorakis S, Karageorgou I, Batra G, Kaltsa O,

Bozinou E, et al. Saffron processing wastes as a bioresource of high-value added compounds: Development of a green extraction process for polyphenol recovery using a natural deep eutectic solvent. Antioxidants. 2019;**8**(12):586

[107] Athanasiadis V, Grigorakis S, Lalas S, Makris DP. Highly efficient extraction of antioxidant polyphenols from *Olea europaea* leaves using an eco-friendly glycerol/glycine deep eutectic solvent. Waste and Biomass Valorization. 2018;**9**(11):1985-1992

[108] Alañón ME, Ivanović M, Gómez-Caravaca AM, Arráez-Román D, Segura-Carretero A. Choline chloride derivative-based deep eutectic liquids as novel green alternative solvents for extraction of phenolic compounds from olive leaf. Arabian Journal of Chemistry. 2018. DOI: 10.1016/j.arabjc.2018.01.003

[109] Dedousi M, Mamoudaki V, Grigorakis S, Makris D. Ultrasound-assisted extraction of polyphenolic antioxidants from olive (*Olea europaea*) leaves using a novel glycerol/sodium-potassium tartrate low-transition temperature mixture (LTTM). Environments. 2017;**4**(2):31

[110] Cui Q, Peng X, Yao XH, Wei ZF, Luo M, Wang W, et al. Deep eutectic solvent-based microwave-assisted extraction of genistin, genistein and apigenin from pigeon pea roots. Separation and Purification Technology. 2015;**150**:63-72

[111] Wei Z, Qi X, Li T, Luo M, Wang W, Zu Y, et al. Application of natural deep eutectic solvents for extraction and determination of phenolics in *Cajanus cajan* leaves by ultra performance liquid chromatography. Separation and Purification Technology. 2015;**149**:237-244

[112] Smink D, Kersten SRA, Schuur B. Recovery of lignin from deep eutectic solvents by liquid-liquid extraction. Separation and Purification Technology. 2020;**235**:116127

[113] Routray W, Orsat V. Microwave-assisted extraction of flavonoids: A review. Food and Bioprocess Technology. 2012;**5**:409-424

[114] Picó Y. Ultrasound-assisted extraction for food and environmental samples. TrAC, Trends in Analytical Chemistry. 2013;**43**:84-99

[115] Jeong KM, Lee MS, Nam MW, Zhao J, Jin Y, Lee DK, et al. Tailoring and recycling of deep eutectic solvents as sustainable and efficient extraction media. Journal of Chromatography. A. 2015;**1424**:10-17

[116] Chen Y, Mu T. Application of deep eutectic solvents in biomass pretreatment and conversion. Green Energy & Environment. 2019;**4**:95-115

[117] Chen W, Xue Z, Wang J, Jiang J, Zhao X, Mu T. Investigation on the thermal stability of deep eutectic solvents. Huaxue Xuebao [Acta Chimica Sinica]. 2018;**34**(8):904-911

[118] Hayyan M, Hashim MA, Hayyan A, Al-Saadi MA, AlNashef IM, Mirghani MES, et al. Are deep eutectic solvents benign or toxic? Chemosphere. 2013;**90**:2193-2195

[119] Hayyan M, Hashim MA, Al-Saadi MA, Hayyan A, AlNashef IM, Mirghani MES. Assessment of cytotoxicity and toxicity for phosphonium-based deep eutectic solvents. Chemosphere. 2013;**93**:455-459

[120] Hayyan M, Looi CY, Hayyan A, Wong WF, Hashim MA. In vitro and in vivo toxicity profiling of ammonium-based deep eutectic solvents. PLoS One. 2015;**10**(2):1-18

[121] Juneidi I, Hayyan M, Hashim MA. Evaluation of toxicity and

biodegradability for cholinium-based deep eutectic solvents. RSC Advances. 2015;**5**:83636-83647

[122] Juneidi I, Hayyan M, Mohd AO. Toxicity profile of choline chloride-based deep eutectic solvents for fungi and Cyprinus carpio fish. Environmental Science and Pollution Research. 2016;**23**:7648-7659

[123] Hayyan M, Mbous YP, Looi CY, Wong WF, Hayyan A, Salleh Z, et al. Natural deep eutectic solvents: Cytotoxic profile. Springerplus. 2016;**5**:913

[124] Ahmadi R, Hemmateenejad B, Safavi A, Shojaeifard Z, Mohabbati M, Firuzi O. Assessment of cytotoxicity of choline chloride-based natural deep eutectic solvents against human HEK-293 cells: A QSAR analysis. Chemosphere. 2018;**209**:831-838

[125] García-Lorenzo A, Tojo E, Tojo J, Teijeira M, Rodríguez-Berrocal FJ, González MP, et al. Cytotoxicity of selected imidazolium-derived ionic liquids in the human Caco-2 cell line. Sub-structural toxicological interpretation through a QSAR study. Green Chemistry. 2008;**10**:508-516

[126] Ranke J, Müller A, Bottin-Weber U, Stock F, Stolte S, Arning J, et al. Lipophilicity parameters for ionic liquid cations and their correlation to in vitro cytotoxicity. Ecotoxicology and Environmental Safety. 2007;**67**(3):430-438

[127] Macário IPE, Jesus F, Pereira JL, Ventura SPM, Gonçalves AMM, Coutinho JAP, et al. Unraveling the ecotoxicity of deep eutectic solvents using the mixture toxicity theory. Chemosphere. 2018;**212**:890-897

[128] Mbous YP, Hayyan M, Wong WF, Looi CY, Hashim MA. Unraveling the cytotoxicity and metabolic pathways of binary natural deep eutectic solvent systems. Scientific Reports. 2017;7(41257):1-14

[129] Radošević K, Čanak I, Panić M, Markov K, Bubalo MC, Frece J, et al. Antimicrobial, cytotoxic and antioxidative evaluation of natural deep eutectic solvents. Environmental Science and Pollution Research. 2018;**25**:14188-14196

Chapter 7

Formulations of BGA for Paddy Crop

Bagampriyal Selvaraj and Sadhana Balasubramanian

Abstract

Blue green algae (BGA) are prokaryotic phototrophic organisms that can fix the atmospheric nitrogen biologically, and were directly applied as a biofertilizers in agricultural fields specifically Paddy field. Since they are having the ability to fix nitrogen, they are formulated with various adsorbents for the purpose of enhancing the crop growth along with maintaining the soil fertility and other soil factors responsible for productivity. The present study revealed that the formulations of blue green algae isolated from paddy fields of southern districts with different adsorbents like alluvial soil, sand, charcoal, and powdered paddy straw. All the adsorbents mixed with blue green algae showed significant growth when compared to the control plant. This determined that the adsorbent formulated mixed blue green algae enhanced the paddy plant growth under greenhouse condition.

Keywords: BGA formulations, adsorbents, cyanobacteria, nitrogen fixers, natural biofertilizers

1. Introduction

Blue green algae are present abundantly in rice fields and are important in helping to maintain rice fields fertility through nitrogen fixation. They are belongs to a group of ubiquitous photosynthetic prokaryotes which possessing the ability to synthesis Chlorophyll a and carryout an important role in nutrient recycling and the maintenance of organic matter in aquatic systems including lakes, rivers and wetland. Nitrogen fixing blue green algae are known to be a prominent component of the microbial population in wetland soils, especially rice fields, contributing significantly to the fertility as a natural bio-fertilizer.

Nitrogen fixation is one of the most important biological processes and, though, the atmosphere contains about 79% nitrogen, most of the plants cannot utilize it. They can utilize combined nitrogen, like ammonium, nitrate, nitrite; etc. This process is called biological nitrogen fixation.

Rice (*Oryza sativa*) is monocot plant, of the grass family (Poaceae). As a cereal grain, it is the most popular cereal worldwide, serving as a stable food for 39 countries and nearly half of the world's population [1]. Globally rice is considered as dietary energy source providing 22% of total energy intake [2]. Rice is second highest worldwide produce and consumed stable food and increasing ratio of population demands more production of rice to meet its consumption [3].

Blue green algal species that thrived in rice field release small quantities of ammonia as the major fertilizing product, and small nitrogenous polypeptides during active growth, whereas most of the fixed products are made available mainly

Figure 1.
Effect of different formulations of mixed blue green algae on paddy plants under greenhouse condition.

through autolysis and decomposition. They have an important role to play in crop production as promising biofertilizers. Here an attempt was made to study the different formulations of blue green algae from the paddy field with the following objectives: Isolation and mass culturing of blue green algae form the areas of selective southern districts of Tamil Nadu. The selective isolated blue green algae have been formulated with different adsorbent like alluvial soil, sand, charcoal, powdered paddy straw and analyzed the interaction effect of various for BGA on vegetative growth of paddy plant (**Figure 1**).

2. Materials and methods

2.1 Sampling

The soil samples collected from the areas namely as Thiruvadanai, of Ramnad, Selugai and Amaravathipudur of Sivagangai and Sakkimangalam of

Figure 2.
Sampling sites.

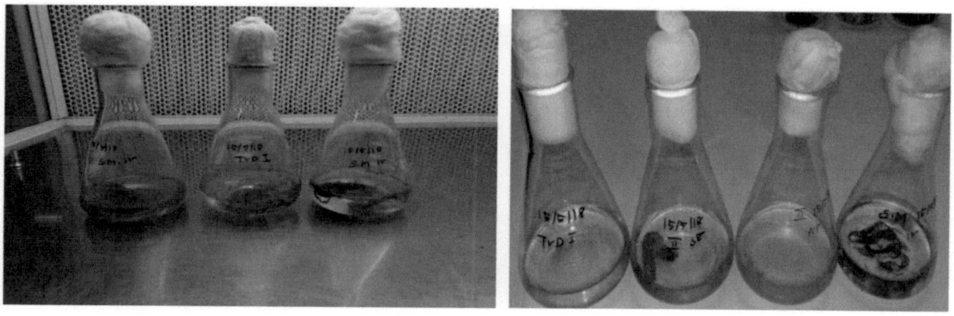

Figure 3.
Sub culturing of isolated blue green algae.

Madurai (**Figure 2**). They were stored at room temperature and were used as samples for further research.

2.2 Culture techniques

The BG_{11} with nitrate and without nitrate medium was prepared and sterilized in autoclave for 121°C, 15 lb pressure for 20 min. After cooling, the samples were inoculated in the BG 11 medium for enrichment. The inoculated flasks were maintained at a temperature of 25°C and 12 h light and 12 h darkness (light intensity 3000 lux).

2.3 Identifying and sub culturing

The blue green algal growth was observed and identifying the organisms under Labomed vision 2000 smart scope B6. The selective identified organisms were sub cultured in BG_0 under lab and maintained for further analysis (**Figure 3**).

2.4 Formulations of BGA

The BGA mixture (10 ml of each *Microcoleus*, *Microcystis*, *Phormidium* and *Gloecapsa*) was added with 50 g of selective adsorbents (alluvial soil, sand, charcoal, powdered paddy straw). Then such combinations were shade dried under laboratory condition. After drying, such mixture was packed in polythene bags further study.

3. Paddy plant selected for general greenhouse procedure

Seeds of Paddy variety CR-1009 were surface sterilized with hot water for 5 min and washed with sterile water repeatedly. Then these seeds were placed in hot water for 10 min to soften the seed coat. Sterile garden soil was used to fill the earthen pots 15 cm height; 52 cm diameter. About 5 kg of sterile soil were taken in each earthen pot which was mixed with different adsorbent formulated BGA. Seeds (15 Nos.) were sown in each pot and germinated seedlings were thinned out to 10 in each pot. The above experimental plants were maintained under greenhouse conditions. The sterilized tap water was used for irrigating the plants. Such experimental pots were assigned for the following treatments:

- C—control (without organism)
- T1—alluvial soil + mixed BGA

- T2—sand + mixed BGA
- T3—charcoal + mixed BGA
- T4—powdered paddy straw + mixed BGA

3.1 Determination of growth

The paddy plant vegetative growth (15th day) was measured with the following growth parameters.

3.2 Determination of fresh and dry weight

The plant materials were cut into bits and weighed. Then they were dried in an oven at 90°C until the weight became constant.

3.3 Shoot and root length determination

The shoot and root lengths of the plants were measured using a meter-scale.

3.4 Determination of leaf number

The number of leaves or leaflets was counted for each plant.

3.5 Estimation of chlorophyll

The experimental leaf tissue was estimated for chlorophyll by following the method of Arnon [4]. Fifty milligram of Leaf tissue was homogenized in 80% pre chilled acetone by using a mortar and pestle and centrifuged at 3000 rpm. The pellet was homogenized again with acetone and was centrifuged repeatedly till the pellet become pale. The collected supernatants were pooled and the absorbance of the supernatant was read at 645 and 663 nm.

The chlorophyll content (mg/g fr.wt) was calculated by using the following formula:

$$\text{Total chlorophyll (mg/g fr.wt) content} = \frac{22.4 \times A645 + 8.02 \times A663}{l \times 1000 \times W} \times V$$

$$\text{Chlorophyll-a (mg/g fr.wt) content} = \frac{22.9 \times A663 - 2.69 \times A645}{l \times 1000 \times W} \times V$$

$$\text{Chlorophyll-b (mg/g fr.wt)} = \frac{22.9 \times A645 - 4.68 \times A663}{l \times 1000 \times W} \times V$$

where l is the path of light length in cm (1 cm), V is the volume of the extract in ml and W is the fresh weight of the sample in g (Chlorophyll contents were expressed either as mg or μg for the plant samples).

3.6 Protein estimation

The experimental fresh leaf tissue of the protein content was estimated by Lowry's method [5]. About 50 mg of the leaf tissue was weighed and was homogenized in hot 80% ethanol and macerate in a mortar with pestle. The supernatant was discarded and the pellet was collected for the analysis purpose. The collected pellet was suspended in a suitable volume of 5% TCA in an ice-bath for 15 min. The pellet was re extracted once in hot absolute ethanol and twice with ethanol-ether mixture, every time discarding the supernatants after centrifugation. Such collected pellet contained proteins and nucleic acids.

The extracted protein sample was placed in 1 ml of sodium hydroxide at 100°C for 4–5 min. The alkaline copper reagent (5 ml) was added and allowed to stand at room temperature for 10 min. Then the folin phenol reagent (0.5 ml) was added rapidly and mixed immediately. After 30 min, the absorbance was measured at 750 nm in a UV–Visible Spectrophotometer. The quantity of protein in the sample was calculated with a standard curve prepared using bovine serum albumin of different concentrations.

3.7 Statistical analysis

The data collected in this study was subjected to statistical methods standard deviation bar charts and pie charts applied [6].

4. Results and discussion

Blue green algae (cyanobacteria) play an important role in maintenance and build-up of soil fertility, consequently increasing rice growth and yield as a natural biofertilizer [7]. They are photosynthetic nitrogen fixers and are free living. Increase in water-holding capacity through their jelly structure [8].

Cyanobacteria are known to be one of the promising supplements to nitrogenous fertilizer, but the process biological nitrogen fixation, mediated through the enzyme nitrogenase may be inhibited in presence readily available nitrogen source. Supplementation of chemical fertilizer with blue green algae could conserve up to 30% of commercial fertilizer and it is generally believed that the nitrogen fixed by these organisms is made available to the rice plants through exudation or autolysis and microbial decomposition. Onkar et al. [9] in addition to contributing fixed nitrogen and adding organic matter to soil such blue green algae are also known to excrete growth promoting substances, solubilize insoluble phosphates, improve fertilizer use efficiency of crop plants and amend the physical and chemical properties of soils, increasing soil aggregate size, there by correcting soil compaction, reduce oxidizable matter of the soil and narrowing down the C:N ratio [10].

Nitrogen fixing filamentous cyanobacteria occurs in wide range of habitats mainly rice-field ecosystem and agricultural fields [11, 12]. In rice field among photosynthetic aquatic organisms, investigations have been emphasized more on isolation and identification of nitrogen fixing cyanobacterial populations in agro-ecosystems for sustainable agriculture.

Shelf-life of cyanobacteria biofertilizer can be augmented by selecting translucent packing material, dry mixing and paddy straw as a carrier [13]. Conventionally, soil has been used as a carrier for cyanobacterial biofertilizers whereas in one study it was reported that soil based inoculums have proved to be disadvantages due to poor inoculums loading, heavy contamination and its bulky nature [14–16]. Sugar cane waste; rice husk [17] and coconut coir [18] was developed as new carrier material [13]. Field trials conducted using straw based, soil based and multani mitti based BGA biofertilizer and it was reported that multani mitti based biofertilizer gave highest yield followed by straw based and soil based BGA inoculants [19].

In the present study the paddy field soil was collected from four different villages namely as Thiruvadanai of Ramnad, Selugai and Amaravathipudur of Sivagangai and Sakkimangalam of Madurai district and blue green algae were isolated as *Microcoleus*, *Microcystis*, *Phormidium* and *Gloecapsa* (**Figure 4**). These isolates were mixed and formulated in four different adsorbents—alluvial soil, sand, charcoal, and powdered paddy straw. The efficiency of such formulates blue green algae mixture on the morphological and physiological activity of paddy plant (15th day growth) was analyzed (**Table 1**). According to this all the formulated BGA (blue green algae) inoculated paddy plant showed progressive increase in shoot and root length, fresh and dry weight, number of leaves, chlorophyll and protein content when compared to control plant. Among these formulations the alluvial soil + BGA treated plants showed better growth by means of increase in chlorophyll and protein content which indicated that the photosynthetic and metabolic activity was enhanced due to this treatment. Blue green algae formulated with adsorbents influenced the paddy plant growth and also they contributed to improve the nitrogen fertility in soil.

The shoot and root length and fresh and dry weight of the paddy plant treated with alluvial soil + Mixed BGA and powdered paddy straw+ Mixed BGA showed maximum (18.83 ± 0.29; 3.93 ± 0.12 cm and 0.21 ± 0.00; 0.052 ± 0.00 g & 18.13 ± 0.12; 3.87 ± 0.12 cm and 0.209 ± 0.001; 0.051 ± 0.00 g) growth when compared to control (13.17 ± 0.29; 2.13 ± 0.23 cm and 0.17 ± 0.00; 0.043 ± 0.00). The number of leaves in all treated plants including control was more or less same (2 or 3). But the chlorophyll *a, b* and total chlorophyll content was higher in (0.485 ± 0.001; 0.1513 ± 0.001; 0.1803 ± 0.001 μg) alluvial soil + Mixed BGA and charcoal + mixed BGA (0.388 ± 0.001; 0.104 ± 0.001; 0.140 ± 0.000 μg) compared to control plant (0.0313 ± 0.001; 0.0187 ± 0.001; 0.0377 ± 0.001 μg) (**Table 1** and

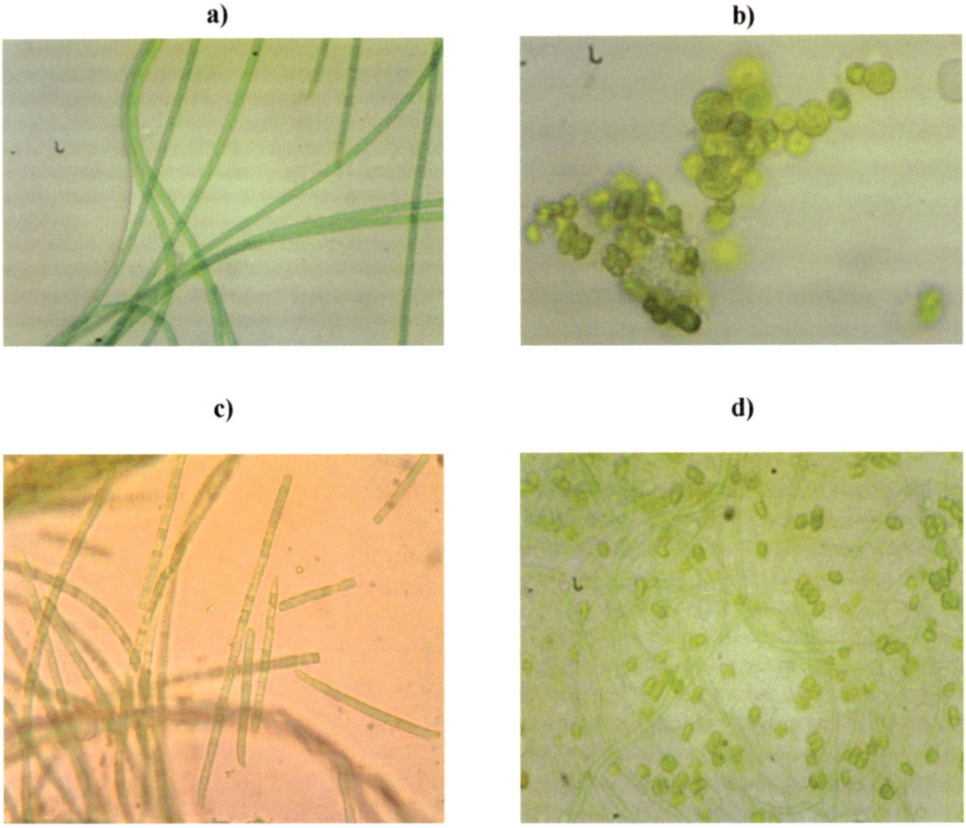

Figure 4.
Microscopic view of isolated blue green algae from soil samples of sampling paddy fields. (a) Microcoleus, *(b)* Microcystis, *(c)* Phormidium, *and (d)* Gloecapsa.

Growth parameters	Control	T1	T2	T3	T4
Shoot length (cm)	13.17 ± 0.29	18.83 ± 0.29	15.5 ± 0.50	17.43 ± 0.40	18.13 ± 0.12
Root length (cm)	2.13 ± 0.23	3.93 ± 0.12	3.37 ± 0.12	3.6 ± 0.10	3.87 ± 0.12
No of leaves	2	3	2	3	3
Fresh weight (g)	0.17 ± 0.00	0.21 ± 0.00	0.20 ± 0.00	0.207 ± 0.00	0.209 ± 0.001
Dry weight (g)	0.043 ± 0.00	0.052 ± 0.00	0.0507 ± 0.00	0.043 ± 0.00	0.051 ± 0.00
Chlorophyll a (μg)	0.0313 ± 0.001	0.485 ± 0.001	0.251 ± 0.002	0.388 ± 0.001	0.279 ± 0.001
Chlorophyll b (μg)	0.0187 ± 0.001	0.1513 ± 0.001	0.074 ± 0.002	0.104 ± 0.001	0.080 ± 0.001

Values are mean of three replicates ± SD.

Table 1.
Effect of different formulations of mixed blue green algae on the growth of Paddy plants under greenhouse condition.

Figure 5). The other formulated BGA treated plants showed minimal chlorophyll contents. The protein content of treated paddy plant with alluvial soil (28%; 2.52 ± 0.02 mg) + Mixed BGA and charcoal + mixed BGA (29%; 2.52 ± 0.00 mg) was significantly maximum when compare the control (10%) paddy plant (0.873 ± 0.06 mg) (**Figure 6**).

Katoh et al. [20] reported that *Nostoc* species are very useful in agricultural applications because of their nitrogen fixation activity, extracellular polysaccharide, photosynthetic system, and particularly desiccation tolerance ability and these properties help to improve the quality of nutrient poor soils. Wetland rice fields could provide an ideal condition for the growth of cyanobacteria, fixing 25–30 kg N ha^{-1} $crop^{-1}$, and reducing the use of urea fertilizer in rice culture by 30% [21, 22]. Algalization of BGA in rice cultivation promotes organic forming without usage of chemical fertilizers and production of organic basmati rice has been reported to develop a potential export market in the country [23].

Cyanobacteria also improve soil characteristics by modifying texture size and subsequent aeration and enhancing carbon content and water holding capacity [24]. Such organisms are one of the major components of the nitrogen fixing biomass in paddy fields. The importance of cyanobacteria in agriculture for paddy cultivation

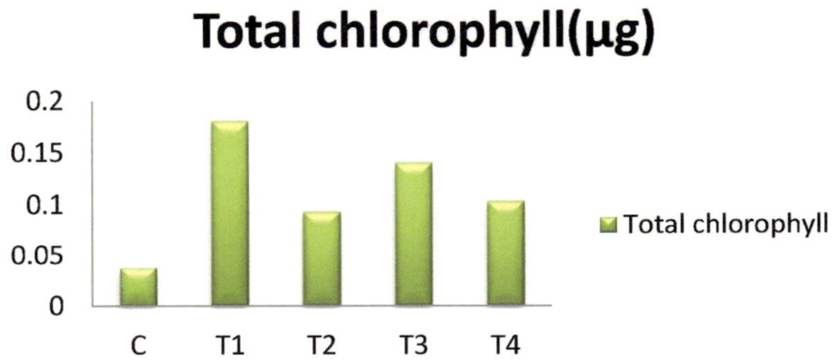

Figure 5.
Effect of different formulations of mixed blue green algae on the total chlorophyll (µg) content of Paddy plants under greenhouse condition. C, control (without organism); T1, alluvial soil + mixed BGA; T2, sand + mixed BGA; T3, charcoal + mixed BGA; T4, powdered paddy straw + mixed BGA.

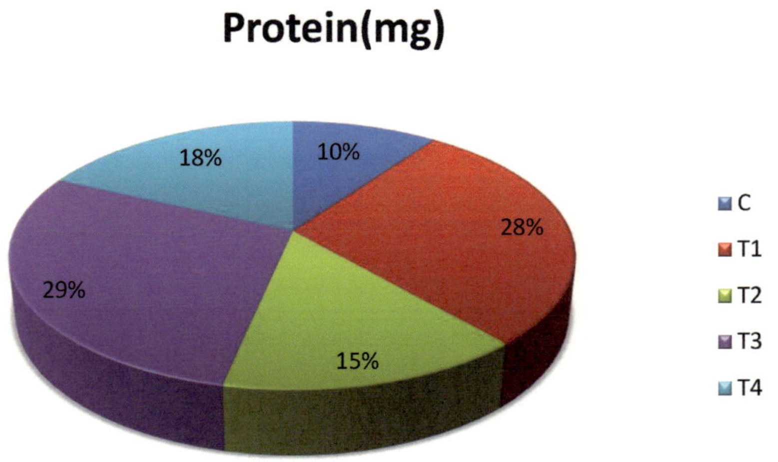

Figure 6.
Effect of different formulations of mixed blue green algae on the protein content (mg) of Paddy plants under greenhouse condition. C, control (without organism); T1, alluvial soil + mixed BGA; T2, sand + mixed BGA; T3, charcoal + mixed BGA; T4, powdered paddy straw + mixed BGA.

is directly proportional to their ability to fix nitrogen and other positive effects for plants and soil. The nitrogen is the second limiting factor next to the water for plant growth in many fields and efficiency of this element is met by fertilizer [25].

Current study suggested that the efficiency of paddy plant growth was enhanced due to the application of formulated BGA with various adsorbents. Such blue green algae were generally applied as biofertilizers in agriculture for improving the soil fertility by the process of biological nitrogen fixation.

5. Conclusion

The blue green algae distributed in different environments. They are actively involved in the fixation of atmospheric nitrogen by the action of nitrogenase enzyme which is present in such organisms but not in plant cells. *Microcoleus*, *Microcystis*, *Phormidium* and *Gloecapsa.* were isolated from the paddy fields of Thiruvadanai, Selugai, Amaravathipudur, Sakkimangalam areas of Ramnad, Sivagangai and Madurai district. The isolated organisms were mass cultured under laboratory condition and mixed well. The BGA mixture formulated with alluvial soil, sand, charcoal and powdered paddy straw were treated on paddy plant showed significant growth compared to control plant. The present study concluded that the alluvial soil and powdered paddy straw formulated BGA promoted the plant growth by means of enhance the morphological growth but chlorophyll and protein content of the alluvial soil and charcoal formulated BGA treated plant showed was maximum. This indicated that the formulated BGA enhanced morphological and photosynthetic efficiency of the paddy plant under greenhouse condition. The application of such bio-mixture in agriculture for crop production not only increase crop yield which may maintain our environment eco-friendly.

Acknowledgements

The authors have expressed their sincere thanks to the President, Vice-President, Secretary, Principal, Head of the Department, Thiagarajar College, Madurai, Tamil Nadu, India for their encouragement, support and provided the necessary facilities for the successful completion of the research work. And also they expressed their sincere thanks to their family and friends for the successful supportive work.

Author details

Bagampriyal Selvaraj[1] and Sadhana Balasubramanian[2]*

1 Thiagarajar College, Madurai, Tamil Nadu, India

2 Post Graduate and Research Department of Botany, Thiagarajar College, Madurai, Tamil Nadu, India

*Address all correspondence to: sadhanakarthik2004@yahoo.co.in

References

[1] Juliana BO. Rice in Human Nutrition. Rome: The International Rice Research Institute and Food Agriculture Organization of the United Nations; 1993

[2] Kainuma K. Rice—Its potential to prevent global hunger. In: Proceedings of the Workshop on Suitable Use of Agricultural Resources and Environment Management with Focus on the Role of Rice Farming. Japan: FAO Association; 2004. pp. 41-46

[3] Kubo M, Purevdorj M. The future of rice production and consumption. Journal of Food Distribution Research. 2004;**35**(1):128-142

[4] Arnon DI. Copper enzymes in isolated chloroplasts. Polyphenol oxidase in *Beta vulgaris*. Plant Physiology. 1949;**24**:1-15

[5] Lowry OH, Rosebrough NJ, Farr AL, Randall RJ. Protein measurement with Folin phenol reagent. The Journal of Biological Chemistry. 1951;**193**:265

[6] Little TM, Hills FC. Agricultural Experimentation. USA: John Wiley and Sons; 1978

[7] Song T, Martensson L, Eriksson T, Zheng W, Rasmussen U. Biodiversity and seasonal variation of cyanobacterial assemblage in a rice paddy field in Fujian, China. The Federation of European Materials Societies Microbiology Ecology. 2005;**54**:131-140

[8] Roger GA, Bergman B, Henriksson E, Udris M. Utilisation of blue green algae as bio fertilizers. Plant and Soil. 1982;**52**(1):99-107

[9] Onkar NT, Tingujam I, Keithellakpam OS, Oinam AS, Gunapathi O, Laxmipriya K, et al. Enumeration, pigments analysis and nitrogenise activity of Cyanobacteria isolated from unexplored rice fields of Manipur, India falling under Indo-Burma biodiversity hotspots. International Journal of Current Microbiology and Applied Sciences. 2015;**4**(6):666-680

[10] Swarnalakshmi K, Dhar DW, Singh PK. Evaluation of blue green algal inoculation on specific soil parameters. Acta Agronomica Hungarica. 2007;**55**(3):307-313

[11] Chunleuchanon S, Sooksawang A, Teacemroong N, Boonkered N. Diversity of nitrogen fixing cyanobacteria under various ecosystems of Thailand population dynamics as affected by environmental factors. World Journal of Microbiology and Biotechnology. 2003;**19**:167-173

[12] Dhar DW, Prasanna R, Singh BV. Comparative performance of three carrier based blue green algal biofertilizerfor sustainable rice cultivation. Journal of Sustainable Agriculture. 2007;**30**(2):41-50

[13] Jha MN, Prasad AN. Useful carrier for cyanobacteria their response to cyanobacterial growth, acetylene-reductase activity, cyanobacterial grazers and paddy yield in calcareous soil. World Journal of Microbiology. 2005;**21**:1521-1527

[14] Bisoyi RN, Singh PK. Effect of phosphorus fertilization on blue-green algal inoculum production and nitrogen yield under field condition. Biology and Fertility of Soils. 1988;**5**:338-343

[15] Jha MN, Prasad AN, Sharma SG, Bharati RC. Effects of fertilization rate and crop rotation on diazotrophic cyanobacteria in paddy field. World Journal of Microbiology and Biotechnology. 2004;**17**:463-468

[16] Raynand PA, Metting B. Colonization potential of cyanobacteria

on temperate Irrigated soils of Washington State, USA. Biological Agriculture and Horticulture. 1988;**5**(3):197-208

[17] Kannaiyan S. Biological fertilizer for sustainable rice (*Oryza sativa*) production. Advances in Agricultural Research in India. 2000;**13**:67-107

[18] Malliga P, Uma L, Subramanian G. Lignolytic activity of the cyanobacterium *Anabaena azollae* ML 2 and the value of coir waste as a carrier for BGA biofertilizer. Microbios. 1996;**86**:175-183

[19] Pabbi S. Cyanobacteria biofertiliser (Reveiw). Journal of Eco-friendly Agriculture. 2008;**3**(2):95-111

[20] Katoh H, Furukawa J, Yokotani KT, Yasuaki K. Isolation and purification of an axenic diazotrophic drought tolerant cyanobacterium, *Nostoc commune*, from natural cyanobacteria l crusts and its utilization for field research on soils polluted with radioisotopes. Biochimica et Biophysica Acta (BBA)-Bioenergetics. 2012;**1817**(8):1499-1505

[21] Choudhary ATMA, Kecskes ML, Kennedy LR. Utilization of BNF technology supplementing urea N for sustainable rice production. Journal of Plant Nutrition. 2004;**37**(10):1627-1647

[22] Hashem MA. Problems and prospects of cyanobacterial biofertilizer for rice cultivation. Australian Journal of Plant Physiology. 2001;**28**(9):881-888

[23] Mulbry W, Kondrad S, Pizarro C. Biofertilizers from algal treatment of dairy and swine manure effluents. Journal of Vegetation Science. 2008;**12**(4):107-125

[24] Richert L, Golubic S, Guedes RL, Ratiskol J, Payri C, Guenennec J. Characterisation of exopolysaccharides produced by Cyanobacteria isolated from Polynesian microbial mats. Current Microbiology. 2006;**51**(6):379-384

[25] Malik FR, Ahmed S, Rizki YM. Utilization of lignocellulosic waste for the nitrogenous biofertilizer. Pakistan Journal of Biological Sciences. 2001;**4**:1217-1220